CREATIVITY IN HUMAN EVOLUTION AND PREHISTORY

THEORETICAL ARCHAEOLOGY GROUP (TAG)

In this series:

The Archaeology of Human Ancestry
Power, Sex and Tradition
edited by James Steele and Stephen Shennan

The Archaeology of Iberia
The Dynamics of Change
edited by Margarita Díaz-Andreu and Simon Keay

Cultural Identity and Archaeology
The Construction of European Communities
edited by Paul Graves-Brown, Siân Jones and Clive Gamble

The Cultural Life of Images
Visual Representation in Archaeology
edited by Brian Leigh Molyneaux

Managing Archaeology
edited by Malcolm A. Cooper, Antony Firth, John Carman and David Wheatley

Theory in Archaeology
A World Perspective
edited by Peter J. Ucko

Time, Tradition and Society in Greek Archaeology
Bridging the 'Great Divide'
edited by Nigel Spencer

CREATIVITY IN HUMAN EVOLUTION AND PREHISTORY

edited by
Steven Mithen

ROUTLEDGE

London and New York

First published 1998
by Routledge
11 New Fetter Lane, London EC4P 4EE

Simultaneously published in the USA and Canada
by Routledge
29 West 35th Street, New York, NY 10001

Chapter 3, 'What is creativity?', was previously published in M. Boden (ed.),
Dimensions of Creativity, 1994, MIT Press.
Reprinted by permission of David Higham Associates.

Typeset in Bembo by The Florence Group, Stoodleigh, Devon
Printed and bound in Great Britain by
Biddles Ltd, Guildford and King's Lynn

British Library Cataloguing in Publication Data
A catalogue record for this book is available from the British Library.

Library of Congress Cataloging in Publication Data
Creativity in human evolution and prehistory /
edited by Steven Mithen.
Includes bibliographical references and index.
1. Human evolution. 1. Creativity ability. 3. Genetic psychology.
4. Prehistoric peoples. I. Mithen, Steven J.
GN281.4.C74 1998
599.93'8–dc21 97–43350

ISBN 0–415–16096–0

Contents

Illustrations

FIGURES

TABLE

CONTRIBUTORS

Margaret A. Boden, School of Cognitive and Computing Sciences, University of Sussex

Richard Bradley, Department of Archaeology, University of Reading

Richard Byrne, School of Psychology, University of St Andrews

Ian Hodder, Department of Archaeology, Cambridge University

Steven L. Kuhn, Department of Anthropology, University of Arizona

Mark Lake, Department of Archaeology, University of Reading

Robert Layton, Department of Anthropology, University of Durham

Caroline Malone, New Hall College, Cambridge University

Steven Mithen, Department of Archaeology, University of Reading

Rick J. Schulting, Department of Archaeology, University of Reading

Mary C. Stiner, Department of Anthropology, University of Arizona

Simon Stoddart, Department of Archaeology, Cambridge University

Colin Renfrew, McDonald Institute for Archaeological Research, Downing Street, Cambridge

CHAPTER ONE

INTRODUCTION

The archaeological study of human creativity

STEVEN MITHEN

To be a good archaeologist one needs to maintain a childlike wonder at those objects we dig up from the ground and the monuments to which we journey. As soon as one loses that sense of awe at the achievements of past individuals and communities, one is on the slippery slope to an academic blackhole, where doing archaeology is no different from doing accountancy. How could past people come up with their ideas – if indeed that is what they did – about the shape of stone tools, the design of cave paintings and the burial of their dead? And how could those seemingly mundane activities that we take for granted, such as acquiring food, building walls and discarding waste, have been undertaken in such an immense variety of ways, many of which we could not have imagined? A very creative mind seems to be the 'bottom line' answer to such questions: a mind that appears to have no bounds in what can be conceived and achieved and which lies at the root of the cultural diversity and change that is so evident from the world around us, let alone the 2.5 million years of the archaeological record.

Archaeologists see more of cultural diversity than those who restrict themselves to the present. They also take upon themselves the burden of explaining this diversity. So surely it is archaeologists who have the most vested interest in understanding this phenomenon of creative thought. Without such understanding, we will never get beyond that crucial, yet so limiting, childlike wonder about the past. This may lead us to gasp at new discoveries, such as the 30,000-year-old paintings from Chauvet cave (Chauvet *et al.* 1996), or the 400,000-year-old wooden javelins from Schöningen (Thieme 1997), and then set us to work at unravelling the economic and social contexts in which these were made. But it also constrains by allowing us to invoke 'the creative mind' as an explanation, without seriously considering what this might mean.

What after all is creativity? Is this a special type of thinking? And if so, is it the preserve of geniuses alone or one that we all share? Or perhaps creativity is an integral part of ordinary thought. Does the term 'creativity' have any value in our attempts to understand the past, or is it just a distraction? Perhaps it is something worse – an imposition of our modern-day values on to the past. For the notion of 'creativity' appears to be extraordinarily valued today in all fields. Hence artists can be forgiven for their lack of drawing skills in light of the creativity they show, while business invests vast sums in management seminars about how to increase creative thought. In our obsession with creativity are we simply attempting to naturalise the idea by claiming its universal existence among all societies in space and time, and hence support the values of our own particular society? And in so doing, are we simply writing the present into the past?

UNDERSTANDING CREATIVITY

Where can those of us whose time is spent with the objects made by long-dead people acquire the understanding of human creativity that we may intuitively feel we need if we are going to achieve satisfactory interpretations of those objects? We might try consulting a spate of recent books by philosophers and psychologists that have addressed this issue; books with titles such as *The Nature of Creativity* (Sternberg [ed.] 1988), *The Creative Mind: Myths and Mechanisms* (Boden 1990), *Creativity: Beyond the Myth of Genius* (Weisberg 1993) or *Dimensions of Creativity* (Boden [ed.] 1994a). Or we may consult less recent books, as trying to unravel the nature of creative thought has been a constant concern in those disciplines. So one might also try *The Act of Creation* (Koestler 1964) or *Creative Person and Creative Process* (Barron 1969).

This strategy of looking into other disciplines was certainly the one I adopted during the last decade as I considered various aspects of early prehistory that seemed to require reference to creative thought for adequate explanation, such as Palaeolithic art and Mesolithic foraging (Mithen 1990, 1996). But as a strategy it has not been wholly successful. Although there is much of value for archaeologists in these works, and particularly in those by Boden (1990, 1994a), so much of the writing by philosophers and psychologists about creativity seems irrelevant to the issues that archaeologists need to address, and the type of data they have available.

The dominant theme within these works is how particular individuals arrived at what, with hindsight, were designated as creative thoughts (in addition to those referenced above, one might consider Gardner 1993, 1994; Simonton 1984; Weisberg 1993). These are predominantly studies of artists and scientists, and in some cases politicians, working in the late nineteenth and early twentieth centuries. Some examples, such as Kekulé's dreams about phantom snakes (which, he claims, resulted in the discovery of the molecular structure of benzene), recur to the point of tedium. Other key subjects include Picasso, Darwin, and Watson and Crick.

Studies of this type seem of little value to archaeologists for several reasons. First and foremost is the focus on particular individuals. Those individuals may be the ones who have been designated as particularly creative in their work, such as Picasso or Einstein, or they may be anonymous individuals in light of the trend towards seeing creative thought as an element of human thought in general and not necessarily linked to the phenomenon of 'genius'. Now while the thoughts and actions of individuals play a critical role in archaeological theory (Shennan 1986; Mithen 1993), particular individuals in prehistory are inaccessible to us, except in extraordinary circumstances. While we may believe that reference to individuals is a vital element of archaeological explanation, we cannot identify particular individuals and analyse why they, rather than others, were responsible for key innovations that influenced the course of prehistory. So, in light of the overwhelming focus on the thought processes of particular individuals in the literature that seeks to understand creativity, so much of that literature is unhelpful to our task as archaeologists.

Even within those periods of study in which a focus on individuals is feasible, there is a growing realisation that to do so imposes a severe constraint on our understanding. Schaffer (1994) has challenged the 'heroic model' of discovery. As he explains, when one removes the myths about single inspirational moments and flashes of genius by past artists and scientists, one finds that 'discovery starts to look less individual and specific, and more like a lengthy process of hard work and negotiation within a set of complex social networks' (1994: 16). Yet the ease with which one can slip back into a biographical mode of analysis for invention and discovery, rather than tackling these complex social networks, means that in the majority of studies particular individuals have maintained their positions of paramount importance. Exceptions do exist. Martindale (1994), for instance, has attempted to devise quantitative measures for the creative output of an entire society and then explored how this varies through time. Nevertheless, such studies are the exception and it is one of the valuable attributes of archaeology that we are forced out of the 'heroic' mode of thinking – as we cannot trace specific individuals, we are forced to focus on those 'complex social networks'.

But the social networks that archaeologists consider are inevitably different to those alluded to by Schaffer (1994). He could embed Faraday's discoveries in the field of electromagnetism into the particular social network of the Royal Institution of the 1840s and monitor how Faraday was elevated into a 'cultural hero' by his contemporaries. Archaeologists consider social networks at a much coarser spatial and chronological scale.

The degree of time resolution that archaeologists can achieve is obviously crucial. The existing literature on creative thought relates not only to particular individuals, but often to a very short period in their lives

– the years immediately preceding Picasso's painting of *Les Demoiselles d'Avignon* in 1907, or Watson and Crick's discovery of the structure of DNA in 1953. Well, even with the remarkable developments in absolute dating methods of the last decade, prehistorians are never going to be able to examine cultural developments over a matter of years. It is unlikely that the first order standard deviation on AMS radiocarbon dates will fall below ± 50 years. As a consequence, much of the existing discussion about how particular social contexts influenced the emergence of apparently creative thinking is as irrelevant to archaeologists, as is the focus on individuals.

Reference to understanding the social context of creative thought leads to a further inadequacy with the current literature on creativity for our needs as prehistorians. Almost without exception, this literature concerns creative thought in western capitalist society during the late nineteenth and early twentieth centuries. Even though capitalist society is wholly anomolous in terms of the span of human history and prehistory, generalisations are attempted from this basis to the general nature of creative thought. A prehistorian is unlikely to find much of value in this literature, as the behaviour they study arises from fundamentally different social contexts and the nature of human cognition cannot be divorced from those contexts (Hodder, in Renfrew *et al.* 1993). Even with the slightest knowledge of the diversity of human societies within which people have lived, the enterprise of finding generalisations about human thought processes from a reliance on recent western society is an enterprise doomed to failure.

Just as the social context that psychologists have studied to understand creativity is too narrow, so too is the biological context. If early prehistorians are asked about the nature of human thought, one of their first questions will be 'what type of human?' – the restriction of the term human to our own species alone, *H. sapiens sapiens*, is a practice adopted by very few archaeologists in light of the great anatomical and cultural similarities between *H. sapiens sapiens* and other members of the *Homo* genus. But is creative thought among anatomically modern humans the same as that among Early Humans, such as the Neanderthals or *Homo erectus*? Is there a difference between creative thought of anatomically modern humans before and after what appears to be an explosion of symbolic behaviour that occurred 35,000 years ago? Not surprisingly, those psychologists and philosophers who have addressed the nature of creativity have not asked such questions, let alone considered the existence of creative thinking among our closest living relatives, the great apes. What is surprising, however, is that they appear not even to have recognised that these are critical questions to address if we are to understand the nature of creativity – all aspects of human cognition can be fully understood only when firmly embedded into an evolutionary context.

So it is readily apparent that the existing literature on creative thinking has limited value to archaeologists. We deal with time frames, social contexts and often biological species that are incompatible with existing theories about how creativity arises and can be explained. But this is not a 'problem' with our discipline. Quite the reverse. It is a problem with the existing literature on creative thought. This needs to be broadened if we are to gain a general understanding of creative thought in humans. This literature needs to escape from its obsession with late nineteenth- and twentieth-century western society and particular individuals within that society. Similarly, it needs to adopt a much greater time depth to its studies so that creative thinking can be examined in an evolutionary context. A hint of recognition exists in the current literature. Hence Boden (1994a and this volume) asks 'how is analogical thinking possible', after recognising that this may lie at the heart of creative thought. A complete answer to this requires an evolutionary perspective, yet none emerges.

Indeed it appears quite remarkable that the immense literature on creative thinking has drawn so little on the data from the archaeological record. Yet this reflects a problem that appears widespread among psychologists and philosophers, and even some anthropologists. For instance, Boyd and Richerson (1996) published a theory regarding the relationship between accumulative cultural change and imitation that is easily falsified by the merest acquaintance with the Palaeolithic archaeological record (Mithen in press). Paul Bloom (in press), writing about the evolution of language, states that 'It turns out that there are data from linguistics, psychology and neuroscience that bear directly on the question of how language evolved' – what about archaeology and palaeoanthropology? Even two of the most prestigious philosophers of mind, Daniel Dennett (1996) and Jerry Fodor (1996), who in a recent exchange of views in the journal *Mind and Language* fundamentally disagreed with each other regarding the evolution of the mind, find that they can agree that this issue cannot be addressed 'until the data is in'. Well, if they took a moment to examine an archaeological textbook they would see that much of the data is not only 'in', but it has been in for many years and subject to immense analysis, interpretation and discussion by archaeologists. In general, the archaeological record appears to be largely ignored by those wishing to make generalisations about the nature of human thought or culture change, and archaeologists should not let this continue. Why? Because we know that without the perspectives offered by the archaeological record, one will only ever get a narrow, biased view of humankind. And it follows that, without examining the prehistory of creative thought, one will only ever get a narrow and biased view of the nature of human creativity.

ARCHAEOLOGICAL STUDIES OF INNOVATION AND CREATIVITY

I have so far identified two reasons why this book, and others like it, are required. First, because, as archaeologists, we regularly find particular artefacts or deal with societies and cultural developments that appear to reflect periods of particularly creative thought in prehistory. Many archaeologists regularly invoke the word 'creative' as either a description or explanation, epitomised in the 'creative explosion' of the Upper Palaeolithic. Yet, as far as I am aware, there has been very little thought by archaeologists as to what this word actually means and to whether the concept of creativity helps us to understand and interpret the past. So I believe that archaeologists need to apply their minds to this issue. Second, as archaeologists are concerned not only with understanding the past, but also with much broader issues about the nature of culture change and of being human, I believe that an archaeological perspective on creativity will contribute to the emergence of a general understanding of the phenomenon of creative thought.

There has, of course, been some important previous work by archaeologists on the issue of creativity, although this has largely been in terms of the closely related theme of innovation. Indeed, while I have traced the need for this book by looking at the broad academic arena in terms of research into creativity, it is also important to see where this book sits in the development of archaeological thought alone. Perhaps its root lies with a paper written two decades ago by Colin Renfrew (1978) entitled 'The anatomy of innovation'. This was a paper that both confused and stimulated me as an undergraduate, but one that asked the same basic questions that lie behind this volume – where do new ideas come from and how does the social context influence their adoption and dissemination?

Innovation was also the theme of a 1989 book edited by Sander van der Leeuw and Robin Torrence entitled *What's New?* In their introductory chapter the editors recognised a problem that continues to exist today: 'creativity has generally been studied in terms of the processes which take place to bring together the ingredients of an idea with very little emphasis on external factors' (Torrence and van der Leeuw 1989: 6). By external factors, they are implicitly referring to the social and economic contexts in which new innovations arise. It was notable, however, that those studies within *What's New?* that addressed creative thought most directly were either ethnographic, such as Rabey's (1989) study of innovation in contemporary Andean technology, or purely theoretical and speculative. Peter Allen's exposition of the claimed new evolutionary synthesis of self-organising systems proposed that 'creativity and change find a place together with structure and function in a new scientific paradigm' (Allen 1989: 277). Such statements are of little value when archaeological studies that explicitly address creativity remain lacking from the discipline. Other references to creativity were also of limited value as their meaning was opaque. Consider McGlade and McGlade (1989: 281), for instance:

> The key element in this process [of innovation] is creativity, which is regarded . . . not as the specialist preserve of a particular group of individuals in society but rather as representing the sum total of human potentiality – in a sense a latent species of 'noise' which may be triggered by a wide diversity of social and cultural circumstances.

As archaeologists we need to do better than this.

One of the questions Torrence and van der Leeuw asked in their introductory chapter in *What's New?* was why, in spite of a widespread belief that innovation is a universal human trait that is almost limitless, 'has this creativity previously been so underplayed in anthropology and archaeology?' (Torrence and van der Leeuw 1989: 4). Two possible reasons were proposed: first, that only 'successful' innovations have ever been thought worthy of study, and consequently those innovations that either failed or did not become widely spread have been ignored; second, that success was further defined in terms of progress. A consequence of this is that changes in the functions of objects have been accorded much greater significance than changes in style alone.

Creativity does not appear to have been accorded any more attention since *What's New*? was published than before, and I suspect that there are more profound reasons for it being so underplayed by archaeologists than those that Torrence and van der Leeuw proposed. Sometimes it appears that every opportunity to avoid the issue is taken. A good example of this is how the notion of 'entoptic' images, which arise when people are in altered states of consciousness, have suddenly become the explanatory factor in so much of prehistoric rock art. Entoptic images consisting of spirals, grids and dots are supposedly universally seen by people when they enter an altered state of consciousness due to features of the human nervous system. After Lewis-Williams and Dowson (1988) argued that entoptic imagery played a significant role in the rock art of the San, almost any occurrence of spirals, abstracts or imaginary beasts in Upper Palaeolithic art (Lewis-Williams 1991; Lorblanchet and Sieveking 1997) and later prehistoric rock art (e.g. Bradley 1989; Dronfield 1995, 1996) have been claimed to be entoptic imagery. Even imagery on Iron Age coins has been interpeted in this fashion (Creighton 1995).

Now it is clear that much creative art during the twentieth century was prompted by the use of drugs to place the artists in altered states of consciousness, and that the use of drugs is ethnographically widespread and they have been widely used in historical periods and most likely throughout prehistory. But to use entoptic imagery as an 'off the peg' explanation for extremely complex designs in prehistoric art simply enables archaeologists to avoid asking questions about the human imagination, creative thought and the symbolism of prehistoric art. When reading such claims, I am reminded of how Salvador Dali (1970: 97) claimed that he had never taken drugs to create his paintings because he *was* the drug –

> why should Dali use drugs when he has discovered that our world is a world of people with hallucinations, where theories, like that of relativity, add to the three dimensions of space and a fourth which is time, the most surrealist and the most hallucinatory of spatial dimensions.

Indeed Timothy Leary, the prophet of LSD, described Dali as the only painter of LSD without LSD (Dali 1970). By such an uncritical use of ideas about entoptic imagery, archaeologists are in effect finding an excuse for not addressing issues about the remarkable creative features of the human mind when in an *un*altered state of consciousness.

CREATIVITY AND COGNITIVE ARCHAEOLOGY

While this book hopes to build upon the previous ideas of Renfrew and those in *What's New?* about innovation, I also view it as a contribution to the development of a mature cognitive archaeology. The term 'cognitive archaeology' was introduced during the early 1980s to refer to studies of past societies in which explicit attention is paid to processes of human thought and symbolic behaviour. Quite how this can be done remains unclear, and a diversity of approaches and studies falls under the poorly defined umbrella of cognitive archaeology (see Renfrew *et al.* 1993). These can be grouped into three broad categories, none of which has explicitly tackled the issue of creative thought.

The first are those studies that began in the late 1970s and that not only laid emphasis on the symbolic aspects of human behaviour but also adopted a post-modernist agenda in which processes of hypothesis testing as a means of securing knowledge were replaced by hermeneutic interpretation (e.g. Hodder 1982, 1986). As such, these studies began as a reaction against what was perceived, largely correctly, as a crude functionalism that had come to dominate archaeological theory, and then attempted to provide a new academic agenda for the discipline epitomised in the volume by Mike Shanks and Chris Tilley (1987) entitled *Re-Constructing Archaeology*. While the critique of functionalism was warmly received and has had a long-lasting effect, it was soon recognised that the epistemology of relativism, the lack of explicit methodology and the refusal to provide criteria to judge between competing interpretations in the more extreme of these studies constituted an appalling academic agenda. Consequently, this type of cognitive archaeology now has a marginal place within the discipline.

A contrasting type of cognitive archaeology has attempted to provide an equal emphasis on symbolic thought and ideology, but sought to do this within a scientific frame of reference in which claims about past beliefs and ways of thought can be objectively evaluated. As such, this archaeology has been characterised as a 'cognitive-processual' archaeology by Colin Renfrew (Renfrew and Bahn 1991) and is broadly the academic context in which I would situate this book (although individual contributors might object, and I dislike the term 'cognitive-processual').

Cognitive-processual archaeology covers an extremely broad range of studies that can themselves be broken into two concerns. One is about cognition as representation and has involved studies of ideology, religious thought and cosmology (e.g. Flannery and Marcus 1983; Renfrew 1985; Renfrew and Zubrow 1993). Such studies argue that these aspects of human behaviour and thought are as amenable to study as are the traditional subjects of archaeology, such as technology and subsistence, which leave more direct archaeological traces. Of course, when written records are available to supplement the archaeological evidence, reconstructions of past beliefs can be substantially developed (Flannery and Marcus, in Renfrew *et al.* 1993).

The other area of cognitive-processual archaeology is about cognition as information processing. This has seen a focus on human decision making, and argued that explicit reference to individuals is required for adequate explanations of long-term cultural change. Perlès (1992), for instance, has attempted to infer the cognitive processes of prehistoric flint knappers, while Mithen (1990) used computer simulations of individual decision making to examine the hunting behaviour of prehistoric foragers. Another important feature has been an explicit concern with the process of cultural transmission. In such studies, attempts have been made to understand how the processes of social learning are influenced by different forms of environment and social organisation (e.g. Mithen 1994; Shennan 1996). More generally, it is argued that the long-term patterns of culture change in the archaeological record, such as the introduction, spread and then demise of particular artefact types (e.g. forms of axe head), can be explained only by understanding both the conscious and unconscious processes of social learning (Shennan 1989, 1991). These two areas of cognitive-processual archaeology – cognition as representation and cognition as information processing – stand rather isolated from each other, and it is hoped that the studies of creativity within this volume help to bridge this divide.

A third category of studies in cognitive archaeology, although one that could be subsumed within 'cognitive-processual archaeology', is those concerned with the evolution of the human mind. As the archaeological record begins 2.5 million years ago with the first stone tools, it covers the period of brain enlargement and the evolution of fully modern language and intelligence. While the fossil record can provide data about brain size, anatomical adaptations for speech, and brain morphology (through the study of endocasts), the archaeological record is an essential means to reconstruct the past thought and behaviour of our ancestors, and the selective pressures for cognitive evolution.

The last decade has seen very substantial developments in this area, although significant contributions had already been made by Wynn (1979, 1981). He attempted to infer the levels of intelligence of human ancestors from the form of early prehistoric stone tools by adopting a recapitualist position and using the developmental stages proposed by Piaget as models for stages of cognitive evolution. While there were other important attempts at inferring the mental characteristics of our extinct ancestors and relatives from their material culture, such as those by Glynn Isaac (1986) and John Gowlett (1984), it was in fact a psychologist, Merlin Donald (1991), who first proposed a theory for cognitive evolution that made significant use of archaeological data in his 1990 book *Origins of the Modern Mind*. His scenario, however, has been challenged by my own work (Mithen 1996), which attempts to integrate current thought in evolutionary psychology with that in cognitive archaeology, and by that of Noble and Davidson (1996), who place greater emphasis on perception and give a more central role to language in the evolution of thought. This evolutionary category of cognitive archaeology has been, perhaps, the most active during the last decade as it has meshed with a general growth of interest in the evolution of the human mind (such as in the emergence of evolutionary psychology).

While these three categories of cognitive archaeology differ in significant ways with regard to both form and content, they also share some over-riding features that form the basis for this volume. The first is that an understanding of human behaviour and society, whether in the distant past or the present, requires explicit reference to human cognition – although there is limited agreement on quite what nature that reference should take. Second, the study of past or present cognition cannot be divorced from the study of society in general – individuals are intimately woven together in shared frames of thought (Hodder, in Renfrew *et al.* 1993). Indeed, the study of past or present minds is hopelessly flawed unless it is integrated into a study of society, economy, technology and environment. Third, material culture is critical not only as an expression of human cognition, but also as a means to attain it. It is evident that the remarkable development of culture since 30,000 years ago, and especially its cumulative character of knowledge (something that had been absent from all previous human cultures), is partly attributable to the disembodiment of mind into material culture – epitomised in the storage of information in paintings and carvings (Donald 1991). In this regard, material culture plays an active role in formulating thought and transmitting ideas, and is not simply a passive reflection of these. Understanding this relationship between material culture and human cognition is one of the key tasks for the future of cognitive archaeology.

This book hopes, therefore, to make a contribution to the field of cognitive archaeology by continuing to struggle with the theoretical and methodological issues of how reference to the human mind can be made in archaeological interpretations – something that is *a priori* taken as a necessity. Human creativity has not been a prominent issue in the post-processual, cognitive-processual or evolutionary categories of cognitive archaeology, the boundaries between which are, of course, extremely blurred. This book is intended to fill that gap and provide a contribution to the emergence of a more mature cognitive archaeology.

THIS VOLUME

I invited the contributors of this volume to present papers at a session at the TAG (Theoretical Archaeology Group) conference at Reading University in December 1995. Several of the invited contributors reacted with a degree of incredulity at my request for papers: "What", they asked, "can I say about creativity?" So to facilitate their own thinking, I circulated the précis of Margaret Boden's 1990 book *The Creative Mind* which had appeared in *Behavioral and Brain Sciences* in 1994 (Boden 1994b). Are the ideas in this paper, I asked, of any value to your own archaeology?

I used this article by Boden as it epitomised to me the strengths and weaknesses of the current literature about creativity. The article, and more generally the book that it summarises, had been of great inspiration in my own work, and I had found Boden's idea that creativity is about the 'transformation and exploration of conceptual spaces' to be most useful. Yet her work lacked an evolutionary perspective and draws solely on case studies of individuals (and computers!) in recent western society. So would those particular ideas about creativity be of value when one's concern is with different time periods, when specific individuals cannot be identified, and when the isolation of activity into discrete domains of 'science' and 'art' is unlikely to have existed? Well, some of the participants in the conference session chose to draw on the paper, as well as on Boden's 1990 book, while others did not. Whether they found Boden's ideas useful or not can be judged from the chapters in the book.

The chapters in this book largely follow the contents and sequence of those papers presented at the TAG meeting of 1995. It will be readily apparent to anyone with the merest acquaintance with archaeological thought during the last two decades that they include chapters by perhaps the three most distinguished theorists working in British archaeology, and who few would deny as having anything other than 'creative minds' when it comes to the practice of archaeology: Richard Bradley, Ian Hodder and Colin Renfrew. With their vast knowledge of the archaeological record, and their varied and at times markedly contrasting attempts at interpretation, these appeared the most important archaeologists to ask about the nature of creative thought in prehistory. Ian Hodder chose to address this issue in light of the long time perspective that the archaeological record provides, while both Richard Bradley and Colin Renfrew tackled specific case studies from later European prehistory.

Four other very distinguished academics were invited and agreed to contribute. Margaret Boden, a cognitive scientist, began the conference session by summarising her views of creative thought, informed as they are by her interests in artificial intelligence. Robert Layton provided a view on creative thought from the perspective of a social anthropologist. He has worked extensively at the interface between archaeology and anthropology, and as author of a seminal book on the anthropology of art (Layton 1981) appeared ideally suited to ease the move from thinking about creativity in modern western society to that in prehistory. Clive Gamble, who during the past decade has produced some of the most innovative ideas in Palaeolithic archaeology, addressed the issue of Neanderthal creativity – which to many may sound like a contradiction in terms. Unfortunately, he was unable to contribute his paper for publication. Richard Byrne, a pioneer in the study of thought among non-human primates, undertook the equally challenging task of examining creative thought by apes and monkeys.

After the conference session, it became clear that a few additional chapters were required for a sufficient examination of creative thought in prehistory. For these I turned to those archaeologists who I consider are producing some of the most exciting current research within the discipline, who specialise in a wide range of chronologial periods and issues and who, I believed, would have something of value to write about creativity – again much to their initial incredulity: Mark Lake, Rick Schulting, Simon Stoddart, Carline Malone, Mary Stiner and Steve Kuhn.

With their contributions, the chapters in this book fell naturally into three groups: 'Perspectives on creativity' (Part I) from a cognitive scientist (Boden), an archaeologist (Hodder) and an anthropologist (Layton); 'The evolution of human creativity' (Part II) with a sequence of chapters, dealing with the common human/ape ancestor (Byrne), the first *Homo* (Lake), the Neanderthals (Kuhn and Stiner) and modern humans (Mithen); and 'Creativity in later prehistoric Europe' (Part III) with contributions addressing issues of burial (Schulting), architecture (Bradley), temple building and figurative art (Malone and Stoddart) and the changing role of the horse in later European society (Renfrew).

It is my impression that all of these authors do maintain an almost childlike wonder at the artefacts, tombs and buildings they deal with, whether these be Oldowan choppers or the wooden wheels of Iron Age carts. But all of them have also attempted to supplement this with an attempt to understand the phenomenon of creative thought with regard to the particular periods and issues with which they are concerned. As such, this book hopes to contribute not only to our understanding of the past, but also to the present, to the nature of human creativity which is central to the nature of being human.

REFERENCES

Allen, P.M. (1989) 'Modelling innovation and change.' In S. van der Leeuw and R. Torrence (eds) *What's New? A Closer Look at the Process of Innovation*, pp. 258–280. London: Unwin Hyman.

Barron, F. (1969) *Creative Person and Creative Process*. New York: Holt, Rinehart & Winston.

Bloom, P. (in press) 'Some issues in the evolution of language and thought.' In D. Cummins and C. Allen (eds) *Evolution of the Mind*. Cambridge, MA: MIT Press.

Boden, M. (1990) *The Creative Mind: Myths and Mechanisms*. London: Weidenfeld & Nicolson.

Boden, M. (ed.) (1994a) *Dimensions of Creativity*. Cambridge, MA: MIT Press.

Boden, M. (1994b) 'Précis of *The Creative Mind: Myths and Mechanisms*.' *Behavioural and Brain Sciences* 17: 519–531.

Boyd, R. and Richerson, P. (1996). 'Why culture is common, but cultural evolution is rare.' In W.G. Runciman, J. Maynard-Smith and R.I.M. Dunbar (eds) *Evolution of Social Behaviour Patterns in Primates and Man*, pp. 77–83. London: Proceedings of the British Academy, vol. 88.

Bradley, R. (1989) 'Deaths and entrances: a contextual analysis of megalithic art.' *Current Anthropology* 30: 68–75.

Chauvet, J.-M., Deschamps, E.B. and Hillaire, C. (1996) *Chauvet Cave: The Discovery of the World's Oldest Paintings*. London: Thames & Hudson.

Creighton, J. (1995) 'Visions of power: imagery and symbols in late Iron Age Britain.' *Britannia* 26: 285–301.

Dali, S. (1970) *Dali by Dali* (translated by E.R. Morse). New York: Abrams Inc.

Dennett, D. (1996) 'Granny versus Mother nature – no contest.' *Mind & Language* 11: 263–269.

D'Errico, F. (1995) 'A new model and its implications for the origin of writing: the la Marche antler revisited.' *Cambridge Archaeological Journal* 5: 163–206.

Donald, M. (1991) *Origins of the Modern Mind*. Cambridge, MA: Harvard University Press.

Dronfield, J. (1995) 'Migraine, light and hallucinogenics: the erocognitive basis of Irish megalithic art.' *Oxford Journal of Archaeology* 14: 261–275.

Dronfield, J. (1996) 'Cognition, art and architecture in Irish paggage-tombs.' *Cambridge Archaeological Journal* 6: 37–72.

Flannery, K.V. and Marcus, J. (1983) *The Cloud People*. New York: Academic Press.

Fodor, J. (1996) 'Deconstructing Dennett's Darwin.' *Mind & Language* 11: 246–262.

Gardner, H. (1993) *The Creators of the Modern Era*. New York: Basic Books.

Gardner, H. (1994) 'The creators' patterns.' In M. Boden (ed.) *Dimensions of Creativity*, pp. 143–158. Cambridge, MA: MIT Press.

Gowlett, J. (1984) 'Mental abilities of early man: a look at some hard evidence.' In R. Foley (ed.) *Hominid Evolution and Community Ecology*, pp. 167–192. London: Academic Press.

Hodder, I. (ed.) (1982) *Symbolic and Structural Archaeology*. Cambridge: Cambridge University Press.

Hodder, I. (1986) *Reading the Past.* Cambridge: Cambridge University Press.

Isaac, G. (1986) 'Foundation stones: early artefacts as indicators of activities and abilities.' In G.N. Bailey and P. Callow (eds) *Stone Age Prehistory*, pp. 221–241. Cambridge: Cambridge University Press.

Koestler, A. (1964) *The Act of Creation.* New York: Macmillan.

Layton, R. (1981) *The Anthropology of Art.* St Albans: Granada.

Lewis-Williams, J.D. (1991) 'Wrestling with analogy: a methodological dilemma in Upper Palaeolithic art research.' *Proceedings of the Prehistoric Society* 57 (1): 149–162.

Lewis-Williams, J.D. and Dowson, T.A. (1988) 'The signs of all times: entoptic phenomena in Upper Palaeolithic art.' *Current Anthropology* 29: 201–245.

Lorblanchet, M. and Sieveking, A. (1997) 'The monsters of Pergouset.' *Cambridge Archaeological Journal* 7: 37–56.

McGlade, J. and McGlade, J.M. (1989) 'Modelling the innovative components of social change.' In S. van der Leeuw and R. Torrence (eds) *What's New? A Closer Look at the Process of Innovation*, pp. 281–299. London: Unwin Hyman.

Marshack, A. (1991) 'The Taï plaque and calendrical notation in the Upper Palaeolithic.' *Cambridge Archaeological Journal* 1: 25–61.

Martindale, C. (1994) 'How can we measure a society's creativity?' In M. Boden (ed.) *Dimensions of Creativity*, pp. 159–197. Cambridge, MA: MIT Press.

Mithen, S. (1990) *Thoughtful Foragers: A Study of Prehistoric Decision Making.* Cambridge: Cambridge University Press.

Mithen, S. (1993) 'Individuals, groups and the Palaeolithic record: a reply to Clark.' *Proceedings of the Prehistoric Society* 59: 393–398.

Mithen, S. (1994) 'Technology and society during the Middle Pleistocene.' *Cambridge Archaeological Journal* 4: 3–33.

Mithen, S. (1996) *The Prehistory of the Mind: A Search for the Origins of Art, Science and Religion.* London and New York: Thames & Hudson.

Mithen, S. (in press) 'Social learning and cultural behaviour: a view from the Old Stone Age.' In H. Box and K. Gibson (eds) *Mammalian Social Learning: Comparative and Ecological Perspectives.* Zoological Society of London Symposium Number 73. Cambridge: Cambridge University Press.

Noble, W. and Davidson. I. (1996) *Human Evolution, Language and Mind: A Psychological and Archaeological Inquiry.* Cambridge: Cambridge University Press.

Perlès, C. (1992) 'In search of lithic strategies: a cognitive approach to prehistoric chipped stone assemblages.' In J.-C. Gardin and C.S. Peebles (eds) *Representations in Archaeology*, pp. 357–384. Bloomington and Indiana: Indiana University Press.

Rabey, M.A. (1989) 'Technological continuity and change among the Andean peasants: opposition between local and global strategies.' In S. van der Leeuw and R. Torrence (eds) *What's New? A Closer Look at the Process of Innovation*, pp. 167–181. London: Unwin Hyman.

Renfrew, C. (1978) 'The anatomy of innovation.' In D. Green, C. Haselgrove and M. Spriggs (eds) *Social Organisation and Settlement*, pp. 89–117. Oxford: British Archaeological Reports, International Series, S47.

Renfrew, C. (1985) *The Archaeology of Cult, the Sanctuary at Phylakopi.* London: Thames & Hudson.

Renfrew, C. and Bahn, P. (1991) *Archaeology: Theories, Methods and Practice.* London: Thames & Hudson.

Renfrew, C. and Zubrow, E. (eds) (1993) *The Ancient Mind*. Cambridge: Cambridge University Press.
Renfrew, C., Peebles, C.S., Hodder, I., Bender, B., Flannery, K.V. and Marcus, J. (1993) 'What is Cognitive Archaeology?' *Cambridge Archaeological Journal* 3: 247–270.
Schaffer, S. (1994) 'Making up discovery.' In M. Boden (ed.) *Dimensions of Creativity*, pp. 13–51. Cambridge, MA: MIT Press.
Shanks, M. and Tilley, C. (1987) *Re-Constructing Archaeology*. Cambridge: Cambridge University Press.
Shennan, S.J. (1986) 'Towards a critical archaeology?' *Proceedings of the Prehistoric Society* 52: 327–356.
Shennan, S.J. (1989) 'Cultural transmission and cultural change.' In S.E. van der Leeuw and R. Torrence (eds) *What's New? A Closer Look at the Process of Innovation*, pp. 330–346. London: Unwin Hyman.
Shennan, S.J. (1991) 'Tradition, rationality and cultural transmission.' In R. Preucel (ed.) *Processual and Postprocessual Archaeologies: Multiple Ways of Knowing the Past*, pp. 197–208. Carbondale: Center for Archaeological Investigations, Southern Illinois University at Carbondale.
Shennan, S.J. (1996) 'Social inequality and the transmission of cultural traditions in forger societies.' In S. Shennan and J. Steele (eds) *The Archaeology of Human Ancestry: Power, Sex and Tradition*, pp. 365–379. London: Routledge.
Simonton, D.K. (1984) *Genius, Creativity, and Leadership*. Cambridge, MA: Harvard University Press.
Sternberg, R.J. (ed.) (1988) *The Nature of Creativity*. New York: Cambridge University Press.
Thieme, H. (1997). 'Lower Palaeolithic hunting spears from Germany.' *Nature* 385: 807–810.
Torrence, R. and van der Leeuw, S. (1989) 'Introduction: What's new about innovation?' In S. van der Leeuw and R. Torrence (eds) *What's New? A Closer Look at the Process of Innovation*, pp. 1–15. London: Unwin Hyman.
van der Leeuw, S. and Torrence, R. (eds) (1989) *What's New? A Closer Look at the Process of Innovation*. London: Unwin Hyman.
Weisberg, R.W. (1993) *Creativity: Beyond the Myth of Genius*. New York: W.H. Freeman.
Wynn, T. (1979) 'The intelligence of later Acheulian hominids.' *Man* 14: 371–391.
Wynn, T. (1981) 'The intelligence of Oldowan hominids.' *Journal of Human Evolution* 10: 529–541.

PART I

PERSPECTIVES ON CREATIVITY

CHAPTER TWO

INTRODUCTION TO PART I

This first part of this volume offers three perspectives on creativity – one by Margaret Boden, a cognitive scientist, one by Ian Hodder, an archaeologist, and one by Robert Layton, an anthropologist. These are markedly different in their treatment of the subject and provide a wide-ranging discussion from which archaeologists of almost any theoretical persuasion will find some material of value.

Boden's work was influential in developing my own interest in creativity and her chapter is reprinted from her own edited collection of papers about the nature of creativity. It is a wide-ranging chapter that explains her idea that creativity can be usefully conceived as the 'exploration and transformation of conceptual spaces'. This is an idea that is used by several of the later contributors in the book and one that should appeal to archaeologists working on a wide range of subjects. Other issues are discussed that also arise within later chapters in the volume, notably the importance of constraining imagination if truly creative ideas are to arise, and the use of material culture to facilitate creative thinking.

With regard to material culture, one of her examples concerns the manner in which jazz composers have developed a special written notation to help keep their various harmonic constraints in mind while making new compositions. But a large part of her chapter is in fact devoted to the use of material culture in facilitating thought about thinking, as she explains how computational approaches – computer programs falling broadly under the remit of artificial intelligence – can help us understand human creativity. This use of computational methods in human psychology is analogous to the manner in which archaeologists can use computer programs to understand past processes of human behaviour. Moreover, as I argue later in the volume, the very first use of symbolic codes by humans is also likely to have been as a means to develop an understanding of human thought processes themselves.

Some archaeologists may question the value of Boden's ideas within this chapter, and their inclusion within a book devoted to prehistory. They might argue that archaeologists should not 'borrow' any more theory from other disciplines – this having been a feature of the discipline for the last 30 years. Others may claim that any computational model of the human mind is necessarily reductionist and consequently of limited value – rather than revealing, they hide the fundamental features of human cognition. It might also be noted that while Boden's work is relevant to the modern western world, its use to archaeologists studying different types of social and economic formations is questionable. Although I have sympathy with all of these points, I believe that they are over-ridden by our need as archaeologists to build bridges to the cognitive sciences. Just as the New Archaeology of the 1960s benefited by drawing on ideas of the New Geography, and the post-processualists of the 1980s drew on sociology to the benefit of the discipline as a whole, archaeologists in the 1990s and beyond need to develop a constructive engagement with various strands of cognitive science. And my judgement is that Boden's ideas, as illustrated by this chapter, have value for archaeologists and her work should be widely read within the discipline.

Ian Hodder's ideas have certainly been widely read within the discipline, as for the last 25 years he has been one of the most – perhaps the most – prolific and important archaeological theorists. His chapter includes some ideas that are similar to Boden's, but it also expands our understanding of creative thinking to a completely new area – creativity as interpretation rather than as problem solving. It is the latter of these that most readily springs to mind when the word creativity is invoked – how an artist solves the problem of representing an emotion, or a scientist solves the problem of seemingly contradictory pieces of evidence. But Hodder argues that we should also conceive of creativity in terms of how people perceive and interpret the world, how they adjust to new events to maintain cultural continuity with the past, rather than to create change. As such this reflects an increasingly wide view within archaeology that explaining stasis in material culture is as challenging and as important as explaining change.

Hodder, however, is not prepared to identify creative thinking as anything readily separable from 'intelligence, imagination, adaptation, agency, problem recognition and problem solving'. And neither is he prepared to identify it as a cognitive process alone: creativity is as much a physical process involving the routines and practices of the body. And certainly one cannot divorce individuals from their social and historical contexts, as he explains how creativity is a social process and people are caught up within the webs of material symbols they create. In all these features, Hodder's approach contrasts with that of Boden; in other respects there are similarities – both see analogy and metaphor as lying at the heart of this thing we designate as creativity.

Robert Layton illustrates the values of ideas contained within both Boden's and Hodder's contributions with his case study of creative thought in traditional Aboriginal society. He illustrates the utility of viewing creative thought in terms of moving around in, exploring, and ultimately transforming conceptual spaces. But he also stresses how creativity is as much to do with interpretation and understanding – echoing Hodder – by arguing that it was the process of making sense of their experiences during colonisation that took indigenous Australians' creativity to its limits. And he suggests that the post-glacial rise in sea level may well have demanded a similar degree of creative response.

These three chapters introduce themes to which many of the later chapters in this book will return. They provide three perspectives on creativity that are, I believe, complementary to each other. From the viewpoint of an archaeologist, they show how cognitive science can provide concepts about creativity of immense value in our task of explaining and/or understanding the past. Yet, as Hodder and Layton show, these are insufficient in themselves and we need a broader view of creativity, one that encompasses the process of interpretation and that identifies creativity as much as a phenomenon of the body, of society and of material culture, as of the mind alone.

CHAPTER THREE

WHAT IS CREATIVITY?

MARGARET A. BODEN

THE DEFINITION OF CREATIVITY

Creativity is a puzzle, a paradox, some say a mystery. Inventors, scientists and artists rarely know how their original ideas arise. They mention intuition, but cannot say how it works. Most psychologists cannot tell us much about it, either. What's more, many people assume that there will never be a scientific theory of creativity – for how could science possibly explain fundamental novelties? As if all this were not daunting enough, the apparent unpredictability of creativity seems to outlaw any systematic explanation, whether scientific or historical.

Why does creativity seem so mysterious? To be sure, artists and scientists typically have their creative ideas unexpectedly, with little if any conscious awareness of how they arose. But the same applies to much of our vision, language and commonsense reasoning. Psychology includes many theories about unconscious processes. Creativity is mysterious for another reason: the very concept is seemingly paradoxical.

If we take seriously the dictionary definition of creation, 'to bring into being or form out of nothing', creativity seems to be not only beyond any scientific understanding, but even impossible. It is hardly surprising, then, that some people have 'explained' it in terms of divine inspiration, and many others in terms of some romantic intuition, or insight. From the psychologist's point of view, however, 'intuition' is the name not of an answer, but of a question. How does intuition work?

People of a scientific cast of mind, anxious to avoid romanticism and obscurantism, generally define creativity in terms of 'novel combinations of old ideas'. Accordingly, the surprise caused by a 'creative' idea is said to be due to the improbability of the combination. Many psychometric tests designed to measure creativity work on this principle.

The novel combinations must be valuable in some way, because to call an idea creative is to say that it is not only new, but interesting. (What is 'interesting' in a given domain is studied, for instance, by literary critics, historians of art and technology, and philosophers of science.) However, combination theorists typically omit value from their definition of creativity. Perhaps they (mistakenly) take it for granted that unusual combinations are always interesting; and perhaps psychometricians make implicit value judgements when scoring the novel combinations produced by their experimental subjects. But since positive evaluation is part of the meaning of 'creative', it should be mentioned explicitly.

Also, combination theorists typically fail to explain how it was possible for the novel combination to come about. They take it for granted, for instance, that we can associate similar ideas and recognise more distant analogies, without asking just how such feats are possible. But in many of the cases that are acclaimed in the history books, it is the recognition of the novel analogy that is so surprising. A psychological theory of creativity needs to explain how analogical thinking works.

These two cavils aside, what is wrong with the combination theory? Many ideas – concepts, theories, instruments, paintings, poems, music – that we regard as creative are indeed based on unusual combinations. For instance, part of the appeal of the Lennon–McCartney arrangement of *Yesterday* was their use of a cello, something normally associated with music of a very different kind; this combination had never happened before. Similarly, the appeal of Heath-Robinson machines lies in the unexpected uses of everyday objects. Again, poets often delight us by juxtaposing seemingly unrelated concepts. For creative ideas such as these, a combination theory (supplemented by a psychological explanation of analogy) would go a long way, and might even suffice.

Many creative ideas, however, are surprising in a deeper way. They concern novel ideas that not only *did not* happen before, but that – in a sense to be clarified below – *could not* have happened before.

Before considering just what this 'could not' means, we must distinguish two senses of *creativity*. One is psychological (let us call it P-creativity), the other historical (H-creativity). A valuable idea is P-creative if the person in whose mind it arises could not have had it before; it does not matter how many times other people have already had the same idea. By contrast, a valuable idea is H-creative if it is P-creative *and* no one else, in all human history, has ever had it before.

H-creativity is something about which we are often mistaken. Historians of science and art are constantly discovering cases in which other people, even in other periods, have had an idea popularly attributed to some national or international hero. Even assuming that the idea was valued at the time by the individual concerned, and by some relevant social group, our knowledge of it is largely accidental. Whether an idea survives, whether it is lost for a while and resurfaces later, and whether historians at a given point in time happen to have evidence of it, depend on a wide variety of unrelated factors. These include fashion, rivalries, illness, trade patterns, economics, war, flood and fire.

It follows that there can be no systematic explanation of H-creativity, no theory that explains *all and only* H-creative ideas. Certainly, there can be no *psychological* explanation of this historical category. But all H-creative ideas, by definition, are P-creative too. So a psychological explanation of P-creativity would include H-creative ideas as well.

Even a psychological explanation of creativity is hostage to the essential element of value. Even a cliché (which may be P-novel to a particular person) can be valued, if it expresses some useful truth; but not all P-novel ideas will be regarded by us (or by the person originating them) as worth having. So a psychologist might sometimes say, 'Certainly, little Ms. Jane Gray could not have had that particular idea before – but it's not worth having, anyway. You can't call it *creative*!' (Likewise, a historian might say, 'Yes, Lady Jane Gray did have that idea before anyone else did – but so what? It's worthless, so you can't call her *creative*!') Such value judgements are to some extent culture-relative, since what is valued by one person or social group may or may not be valued – praised, preserved, promoted – by another (Brannigan 1981).

However, our concern is with the origin of creative ideas, not their valuation (the context of discovery, not of justification). Admittedly, criteria of valuation sometimes enter into the originating process itself, so the distinction is more analytical than psychological. But our prime focus is on how creative ideas can arise in people's minds.

What does it mean to say that an idea 'could not' have arisen before? Unless we know that, we cannot make sense of P-creativity (or H-creativity either), for we cannot distinguish radical novelties from mere 'first-time' newness.

An example of a novelty that clearly *could* have happened before is a newly generated sentence, such as 'The pineapples are in the bathroom cabinet, next to the oil paints that belonged to Machiavelli.' I have never thought of that sentence before, and almost certainly no one else has either.

The linguist Noam Chomsky remarked on this capacity of language speakers to generate first-time novelties endlessly, and he called language 'creative' accordingly. His stress on the infinite fecundity of language was correct, and highly relevant to our topic. But the word 'creative' was ill-chosen. Novel though the sentence about Machiavelli's oil paints is, there is a clear sense in which it *could* have occurred before. For it can be generated by the same rules that can generate other English sentences. Any competent speaker of English could have produced that sentence long ago – and so could a computer, provided with English vocabulary and grammatical rules. To come up with a new sentence, in general, is not to do something P-creative.

The 'coulds' in the previous paragraph are computational 'coulds'. In other words, they concern the set of structures (in this case, English sentences) described and/or produced by one and the same set of generative rules (in this case, English grammar).

There are many sorts of generative system: English grammar is like a mathematical equation, a rhyming schema for sonnets, the rules of chess or tonal harmony, or a computer program. Each of these can (timelessly) describe a certain set of possible structures. And each might be used, at one time or another, in actually producing those structures.

Sometimes we want to know whether a particular structure could, in principle, be described by a specific schema, or set of abstract rules. Is '49' a square number? Is 3,591,471 a prime? Is this a sonnet, and is that a sonata? Is that painting in the Impressionist style? Is that building in the 'prairie house' style? Could that geometrical theorem be proved by Euclid's methods? Is that word string a sentence? Is a benzene ring a molecular structure that is describable by early nineteenth-century chemistry (before Friedrich von Kekulé's famous fireside daydream of 1865)? To ask *whether an idea is creative or not* (as opposed to how it came about) is to ask this sort of question.

But whenever a particular structure is produced in practice, we can also ask what generative processes actually went on in its production. Did a particular geometer prove a particular theorem in this way, or in that? Was the sonata composed by following a textbook on sonata form? Did the architect, consciously or unconsciously, design the house by bearing certain formal principles in mind? Did Kekulé rely on the then-familiar principles of chemistry to generate his seminal idea of the benzene ring, and if not, how did he come up with it? To ask how an idea (creative or otherwise) *actually arose* is to ask this type of question.

We can now distinguish first-time novelty from radical originality. A merely novel idea is one that can be described and/or produced by the same set of generative rules as are other, familiar ideas. A genuinely original or radically creative idea is one that cannot. It follows that the ascription of creativity always involves tacit or explicit reference to some specific generative system.

It follows, too, that constraints – far from being opposed to creativity – make creativity possible. To throw away all constraints would be to destroy the capacity for creative thinking. Random processes alone, if they happen to produce anything interesting at all, can result only in first-time curiosities, not radical surprises. (This is not to deny that, in the context of background constraints, randomness can sometimes contribute to creativity [Boden 1990: ch. 9].)

EXPLORING AND TRANSFORMING CONCEPTUAL SPACES

The definition of creativity given above implies that, with respect to the usual mental processing in the relevant domain (chemistry, poetry, music, etc.), a creative idea is not just improbable, but *impossible*. How could it arise, then, if not by magic? And how can one impossible idea be more surprising, more creative, than another? If the act of creation is not mere combination, or what Arthur Koestler (1964) called 'the bisociation of unrelated matrices', what is it? How can creativity possibly happen?

To understand this, we need the notion of a conceptual space. (This idea is used metaphorically here; later, we shall see how conceptual spaces can be described in specific, rigorous and explicit terms.) The dimensions of a conceptual space are the organising principles that unify and give structure to a given domain of thinking. In other words, it is the generative system that underlies that domain and defines a certain range of possibilities: chess moves, or molecular structures, or jazz melodies.

The limits, contours, pathways and structure of a conceptual space can be mapped by mental representations of it. Such mental maps can be used (not necessarily consciously) to explore – and to change – the spaces concerned.

Conceptual spaces can be explored in various ways. Some exploration merely shows us something about the nature of the relevant conceptual space that we had not explicitly noticed before. When Dickens described Scrooge as 'a squeezing, wrenching, grasping, scraping, clutching, covetous old sinner', he was exploring the space of English grammar. He was reminding the reader (and himself) that the rules of grammar allow us to use any number of adjectives before a noun. Usually, we use only two or three; but we may, if we wish, use seven (or more). That possibility already existed, although its existence may not have been realised by the reader.

Some exploration, by contrast, shows us the limits of the space, and perhaps identifies points at which changes could be made in one dimension or another. One modest example occurred at the Mad Tea-Party:

> 'It's always six o'clock now,' the Hatter said mournfully.
> A bright idea came into Alice's head. 'Is that the reason so many tea-things are put out here?' she asked.
> 'Yes, that's it,' said the Hatter with a sigh: 'it's always tea-time, and we've no time to wash the things between whiles.'
> 'Then you keep moving round, I suppose?' said Alice.
> 'Exactly so,' said the Hatter: 'as the things get used up.'
> 'But what happens when you come to the beginning again?' Alice ventured to ask.
>
> (Carroll 1859: ch. 7)

As usual in Wonderland, Alice got no sensible reply (the March Hare interrupted, saying, 'Suppose we change the subject'). But her question was a good one. She had noticed that the conceptual space of the Mad Tea-Party involved a repetitive procedure (moving from one place setting to the next) that eventually would reach a point where something new would have to happen. That 'something' could be many different things. When there were no clean things left on the tea table, the moving around might stop permanently, and the creatures would go hungry; or it might stop temporarily, while the clock was ignored and the washing up was done; or the creatures might drop their previous qualms about hygiene, and go on using the unwashed plates, which would get dirtier with every cycle; or they might bend down to pick some grass and quickly wipe the dishes with it. . . . The March Hare's interruption prevented Alice from finding out which (if any) of these was chosen. The point, however, is that she had identified a specific limitation of this space, and had asked what could be done to overcome it.

To overcome a limitation in a conceptual space, one must change it in some way. One may also change it, of course, without yet having come up against its limits. A small change (a 'tweak') in a relatively superficial dimension of a conceptual space is like opening a door to an unvisited room in an existing house. A large change (a 'transformation'), especially in a relatively fundamental dimension, is more like the instantaneous construction of a new house, of a kind fundamentally different from (albeit related to) the first. Most of the changes to tea-party behaviour suggested above would be small, allowing the tea party to continue but in a slightly modified form. The first, however, might destroy the space, if the participants starved to death.

A complex example of structural exploration and change can be found in the development of post-Renaissance western music. This music is based on the generative system known as tonal harmony. From its origins to the end of the nineteenth century, the harmonic dimensions of this space were continually tweaked to open up the possibilities (the rooms) implicit in it from the start. Finally, a major transformation generated the deeply unfamiliar (yet closely related) space of atonality.

Each piece of tonal music has a 'home key', from which it starts, from which (at first) it does not stray, and in which it must finish. Reminders and reinforcements of the home key are provided, for instance, by fragments of scales decorating the melody, or by chords and arpeggios within the accompaniment. As time passed, the range of possible home keys became increasingly well defined. Johann Sebastian Bach's 'Forty-Eight', for example, was a set of preludes and fugues specifically designed to explore – and clarify – the tonal range of the well-tempered keys.

But travelling along the path of the home key alone became insufficiently challenging. Modulations between keys were then allowed, within the body of the composition. At first, only a small number of modulations (perhaps only one, followed by its 'cancellation') were tolerated, between strictly limited pairs of harmonically related keys. Over the years, however, the modulations became increasingly daring and increasingly frequent – until in the late nineteenth century there might be many modulations within a single bar, not one of which would have appeared in early tonal music. The range of harmonic relations implicit in the system of tonality gradually became apparent. Harmonies that would have been unacceptable to the early musicians, who focused on the most central or obvious dimensions of the conceptual space, became commonplace.

Moreover, the notion of the home key was undermined. With so many, and so daring, modulations within the piece, a 'home key' could be identified not from the body of the piece, but only from its beginning and end. Inevitably, someone (it happened to be Arnold Schoenberg) eventually suggested that the convention of the home key be dropped altogether, because it no longer made sense in terms of constraining the composition as a whole. (Significantly, Schoenberg suggested various new constraints to structure his music making: using every note in the chromatic scale, for instance.)

Another example of extended exploration, this time with an explicit map to guide it, was the scientific activity spawned by Mendeleyev's periodic table. This table, produced in the 1860s for an introductory chemistry textbook, arranged the elements in rows and columns according to their observable properties and behaviour. All the elements within a given column were in this sense 'similar'. But Mendeleyev left gaps in the table, predicting that unknown elements would eventually be found with the properties appropriate to these gaps (no known element being appropriate).

Sure enough, in 1879 a new element (scandium) was discovered whose properties were what Mendeleyev had predicted. Later, more elements were discovered to fill the other gaps in the table. And later still, the table (based on observable properties) was found to map on to a classification in terms of atomic number. This classification explained why the elements behaved in the systematic ways noted by Mendeleyev.

These examples show that exploration often leads to novel ideas. Indeed, it often leads to ideas, such as new forms of harmonic modulation, that are normally called creative. In that sense, then, conceptual exploration is a form of creativity. However, exploring a conceptual space is one thing: transforming it is another. What is it to transform such a space?

One example has been mentioned already: Schoenberg's dropping the home-key constraint to create the space of atonal music. Dropping a constraint is a general heuristic, or method, for transforming conceptual spaces. The deeper the generative role of the constraint in the system concerned, the greater the transformation of the space.

Non-Euclidean geometry, for instance, resulted from dropping Euclid's fifth axiom, about parallel lines meeting at infinity. (One of the mathematicians responsible was Lobachevsky, immortalised not only in encyclopedias of mathematics but also in the songs of Tom Lehrer.) This transformation was made 'playfully', as a prelude to exploring a geometrical space somewhat different from Euclid's. Only much later did it turn out to be useful in physics.

Another very general way of transforming conceptual spaces is to 'consider the negative': that is, to negate a constraint. (Negating a constraint is not the same as dropping it. Suppose someone gets bored with eating only red sweets: to choose *any non-red sweet* is different from choosing *any sweet, whatever its colour.*)

One well-known instance of constraint negation concerns Kekulé's discovery of the benzene ring. He described it like this:

> I turned my chair to the fire and dozed. Again the atoms were gambolling before my eyes. . . . [My mental eye] could distinguish larger structures, of manifold conformation; long rows, sometimes more closely fitted together; all twining and twisting in snakelike motion. But look! What was that? One of the snakes had seized hold of its own tail, and the form whirled mockingly before my eyes. As if by a flash of lightning I awoke.
>
> Findlay 1965: 38–39

This vision was the origin of his hunch that the benzene molecule might be a ring, a hunch that turned out to be correct.

Prior to this experience, Kekulé had assumed that all organic molecules are based on strings of carbon atoms (he had produced the string theory some years earlier). But for benzene, the valencies of the constituent atoms did not fit.

We can understand how it was possible for him to pass from strings to rings, as plausible chemical structures, if we assume three things (for each of which there is independent psychological evidence). First, that snakes and molecules were already associated in his thinking. Second, that the topological distinction between open and closed curves was present in his mind. And third, that the 'consider the negative' heuristic was present also. Taken together, these three factors could transform 'string' into 'ring'.

A string molecule is what topologists call an open curve. Topology is a form of geometry that studies not size or shape, but neighbour relations. An open curve has at least one end point (with a neighbour on only one side), whereas a closed curve does not. An ant crawling along an open curve can never visit the same point twice, but on a closed curve it will eventually return to its starting point. These curves need not be curvy in shape. A circle, a triangle and a hexagon are all closed curves; a straight line, an arc and a sine wave are all open curves.

If one considers the negative of an open curve, one gets a closed curve. Moreover, a snake biting its tail is *a closed curve that one had expected to be open.* For that reason, it is surprising, even arresting ('But look! What was that?'). Kekulé might have had a similar reaction if he had been out on a country walk and happened to see a snake with its tail in its mouth. But there is no reason to think that he would have been stopped in his tracks by seeing a Victorian child's hoop. A hoop is a hoop, is a hoop: no topological surprises there. (No topological surprises in a snaky sine wave, either: so two intertwined snakes would not have interested Kekulé, though they might have stopped Francis Crick dead in his tracks, a century later.)

Finally, the change from open curves to closed ones is a topological change, which by definition will alter neighbour relations. And Kekulé was an expert chemist, who knew very well that the behaviour of a molecule depends not only on what the constituent atoms are, but also on how they are juxtaposed. A change in atomic neighbour relations is very likely to have some chemical significance. So it is understandable that he had a hunch that this tail-biting snake molecule might contain the answer to this problem.

Plausible though this talk of conceptual spaces may be, it is – thus far – largely metaphorical. I have claimed that in calling an idea creative one should specify the particular set of generative principles with respect to which it is impossible. But I have not said how the (largely tacit) knowledge of literary critics, musicologists and historians of art and science might be explicitly expressed within a psychological theory of creativity. How can this be done? And, the putative structures having been made explicit, how can we be sure that the mental processes specified by the psychologist really are powerful enough to generate such-and-such ideas from such-and-such structures? This is where computational psychology can help us.

THE RELEVANCE OF COMPUTATIONAL PSYCHOLOGY

Computational psychology draws many of its theoretical concepts from artificial intelligence, or AI. Artificial intelligence studies the nature of intelligence in general, and its method is to try to enable computers to do the sorts of things that minds can do: seeing, speaking, story telling, and logical or analogical thinking.

But how can computers have anything to do with creativity? The very idea, it may seem, is absurd. The first person to denounce this apparent absurdity was Ada, Lady Lovelace, the friend and collaborator of Charles Babbage. She realised that Babbage's 'Analytical Engine' – in essence, a design for a digital computer – could in principle 'compose elaborate and scientific pieces of music of any degree of complexity or extent'. But she insisted that the creativity involved in any elaborate pieces of music emanating from the Analytical Engine would have to be credited not to the engine, but to the engineer. As she put it, 'The Analytical Engine has no pretensions whatever to *originate*, anything. It can do [only] *whatever we know how to order it* to perform.'

If Lady Lovelace's remark means merely that *a computer can do only what its program enables it to do*, it is correct – and, from the point of view of theoretical psychology, helpful and important. It means, for instance, that if a program manages to play a Chopin waltz expressively, or to improvise modern jazz, then the musical structures and procedures (the generative structures) in that program *must* be capable of producing those examples of musical expression or improvisation. (It does not follow that human musicians do it in the same way: perhaps there is reason to suspect that they do not. But the program specifies, in detail, *one* way in which such things can be done. Alternative theories, involving different musical structures or psychological processes, should ideally be expressed at a comparable level of detail.)

But if Lady Lovelace's remark is intended as an argument denying any interesting link between computers and creativity, it is too quick and too simple. We must distinguish four different questions, which are often confused with each other. I call them Lovelace questions, because many people would respond to them (with a dismissive 'No!') by using the argument cited above.

The first Lovelace question is whether computational concepts can help us understand how *human* creativity is possible. The second is whether computers (now or in the future) could ever do things that at least *appear to be* creative. The third is whether a computer could ever *appear to recognise* creativity – in poems written by human poets, for instance, or in its own novel ideas about science or mathematics. And the fourth is whether computers themselves could ever *really* be creative (as opposed to merely producing apparently creative performance, whose originality is wholly due to the human programmer).

Our prime interest is in the first Lovelace question, which focuses on the creativity of human beings. The next two Lovelace questions are psychologically interesting insofar as they throw light on the first. For our purposes, the fourth Lovelace question can be ignored. It is not a scientific question, as the others are, but in part a philosophical worry about 'meaning' and in part a disguised request for a moral-political decision (Boden 1990: ch. 11).

The answers I shall propose to the first three questions are, respectively: *Yes, definitely*; *Yes, up to a point;* and *Yes, necessarily (for any program that appears to be creative)*. In short, computational ideas can help us to understand how human creativity is possible. This does not mean that creativity is predictable, nor even that an original idea can be explained in every detail after it has appeared. But we can draw on computational ideas in understanding in scientific terms how 'intuition' works.

The psychology of creativity can benefit from AI and computer science *precisely because* – as Lady Lovelace pointed out – a computer can do only what its program enables it to do. On the one hand, computational concepts, and their disciplined expression in programming terms, help us to specify generative principles clearly. On the other hand, computer modelling helps us to see, in practice, what a particular generative system *can* and *cannot* do.

The results may be surprising, for the generative potential of a program is not always obvious: the computer may do things we did not know we had 'ordered it' to perform. And, all too often, it may fail to do things that we fondly believed we had allowed for in our instructions. So expressing a psychological theory as a program to be run on a computer is an excellent way of testing its clarity, its coherence and its generative potential.

In the discussion so far, I have relied on some computational concepts, such as *generative system* and *heuristic*. I have not had to explain these concepts by reference to computer programs, for they were introduced into our language long before the invention of computers, by people studying the nature and psychology of mathematical proof.

For psychological purposes, however, it can be helpful to ask how specific heuristics might play a role in a functioning computer. For what they do in that artificial context can be clearly understood, and so may help us to clarify what could (and what could not) be going on in human thought. Similarly, a consideration of actual AI programs can help us to understand more clearly how conceptual spaces can be identified, constructed, explored or transformed. Even if a program falls far short of the comparable human reality (which is usually the case), its failings can lead, in true Popperian fashion, to progress in the psychological theory concerned.

CONCEPTUAL SPACES IN THE VISUAL ARTS

Many human artists use computers as tools, to help them create things they could not have created otherwise. A graphics artist, for instance, may get new ideas from computer graphics, and so-called computer music may use sounds that no orchestra could produce. Most of these examples, however, are not pertinent here.

The relevant cases are those where the new ideas are made possible by a systematic analysis of the artistic genre concerned. Most relevant of all, for our purposes, are those (few) computer programs that produce aesthetically valuable creations themselves or which, in their attempts to do so, throw light on the psychological processes underlying human art. The conceptual spaces involved may be highly complex, and the computational power of a computer will then be needed to show just what spatial forms the (programmed) genre can or cannot generate.

Some conceptual spaces involved in spatial design have been mapped as algorithms simple enough to be followed 'by hand', at least if the highest levels of potential complexity are ignored (Stiny 1991; Stiny and Gips 1978). Architects and environmental planners, for instance, have used 'spatial grammars' to generate new 'sentences', novel spatial structures that are intuitively acceptable in stances of the genre concerned. Previously unseen examples of Palladian villas, Mughul gardens and Frank Lloyd Wright's 'prairie houses' have been designed accordingly (Koning and Eizenberg 1981; Stiny and Mitchell 1978, 1980). The decorative arts have received similar attentions: traditional Chinese lattice designs have been described by a computer algorithm that generates the seemingly irregular patterns called 'ice rays' as well as the more obviously regular forms (Stiny 1977).

The ice ray example shows that a rigorous analysis of a conceptual space can uncover hidden regularities, and so increase – not merely codify – our aesthetic understanding of the style. The same applies to the analytical work on prairie houses. The architectural grammarians who developed this analysis point out that a renowned expert on Lloyd Wright's buildings had been unable to explicate the notion of *balance* in prairie-style houses: he had described it as 'occult'. Their analysis has uncovered the principles of spatial balance involved. It shows which aspects are relatively fundamental (like Euclid's axioms in geometry, or *NP* and *VP* in syntax), and how certain features are constrained by others.

In the genre of prairie houses, the origin of the generative design (the first 'axiom') is the fireplace. The majority of these houses have only one fireplace. Occasionally, however, Lloyd Wright replaced the single hearth by several fireplaces. Because of the pivotal role of the fireplace in this particular style, this number variation generates 'a veritable prairie village of distinct but interacting prairie-style designs', all within a single building (Koning and Eizenberg 1981: 322).

The aesthetic styles of Palladian villas, Mughul gardens, prairie houses and Chinese lattices are all relatively austere. So perhaps it is not surprising that 'grammatical' analyses of them can be found. Nor is it surprising that these analyses can often be followed by hand: an architect can design a prairie house, using the relevant grammar, without ever using a computer. What of more 'free' aesthetic styles, and art objects inspired by natural forms rather than by geometrical shapes?

Consider line drawings of human figures, for example. Some computational work done by Harold Cohen – already a well-known professional painter when he started working with computers – is pertinent here. Over the past two decades, Cohen has written a series of programs that produce pleasing, and unpredictable, line drawings (McCorduck 1991). I have one in my office, and on several occasions a visitor has spontaneously remarked, 'I like that drawing! Who did it?' These drawings have been exhibited at the Tate and other major art galleries around the world, and not just for their curiosity value.

Each of Cohen's programs explores a certain style of line drawing and a certain subject matter. The program may draw acrobats with large beach balls, for instance, or human figures in the profuse vegetation of a jungle. (Cohen has recently exhibited a program that colours its own pictures; usually, however, he colours his programs' drawings by hand.)

Much as human artists have to know about the things they are depicting, so each of Cohen's programs needs an internal model of its subject matter. This model is not a physical object, like the articulated wooden dolls found in artists' studios, but a generative system: what one might call a 'body grammar'. It is a set of abstract rules that specify, for instance, not only the anatomy of the human body (two arms, two legs), but also how the various body parts appear from various points of view. An acrobat's arm pointing at the viewer will be foreshortened; a flexed arm will have a bulging biceps; and an arm lying behind another acrobat's body will be invisible.

The program can draw acrobats with only one arm visible (because of occlusion), but it cannot draw one-armed acrobats. Its model of the human body does not allow for the possibility of there being one-armed people. They are, one might say, unimaginable. If, as a matter of fact, the program has never produced a picture showing an acrobat's right wrist occluding another acrobat's left eye, that is a mere accident of its processing history: it *could* have done so at any time. But the fact that it has never drawn a one-armed acrobat has a deeper explanation: such drawings are, in a clear sense, *impossible.*

If Cohen's program were capable of 'dropping' one of the limbs (as a geometer may drop Euclid's fifth axiom, or Schoenberg the notion of the home key), it could then draw one-armed, or one-legged, figures. A host of previously unimaginable possibilities, only a subset of which might ever be actualised, would have sprung into existence at the very moment of dropping the constraint that there must be (say) a left arm.

A superficially similar but fundamentally more powerful transformation might be effected if the numeral '2' had been used in the program to denote the number of arms. For a numeral is a *variable*, in the sense that one numeral may be replaced by another. So '2' can be replaced by '1' – or, for that matter, by '7'. And depending on the role played by the numeral in the relevant computational system, the result might be a superficial or a fundamental change. We have seen, for instance, that a prairie house may have one fireplace or several, and that the basic architectural form of the whole house depends on how many fireplaces there are.

A general purpose tweaking–transformational heuristic might look out for numerals, and try substituting varying values. Kekulé's chemical successors employed such a heuristic when they asked whether any ring molecules could have five atoms in the ring, not six. (They also treated carbon as a variable – as a particular instance of the class of elements – when they asked whether molecular rings might include nitrogen or phosphorus atoms.) A program that (today) drew one-armed acrobats for the first time by employing a 'vary-the-variable' heuristic *could* (tomorrow) be in a position to draw seven-legged acrobats as well. A program that merely 'dropped the left arm' *could not.*

Suppose that Cohen's program (or Cohen himself) were to allow the left arm to be omitted, without making any other change to the program. The resulting pictures might not be so plausible, or so pleasing.

The reason is that the program's current world model contains rules dealing with human stability and picture balance, some of which may implicitly or explicitly assume that all people have four limbs. If so, a three-limbed person (one limb having been 'dropped') might be drawn in a physically impossible bodily attitude. Human artists drawing a one-armed person would not do this, unless they were deliberately contravening the laws of gravity (as in a Chagall dreamscape). Likewise, a one-armed person placed carefully on the page might look *visually* unbalanced if the aesthetic criteria governing that placement currently assume a two-armed person. (This is an example of the fact mentioned earlier: evaluative criteria can enter into the generation of a conceptual structure, as well as into its selection/rejection *post hoc*.)

The psychological interest of Cohen's work is that the constraints – anatomical, physical and aesthetic – written into his programs are perhaps a subset of those that human artists respect when drawing in comparable styles. A host of questions arises about just what those constraints may be. And a host of issues can be explored by building additional or alternative rules into Cohen's programs, and examining the range of structures that result. Such experiments cannot be done by hand without begging the very questions we are interested in. To understand the potential (and some of the limits) of this genre clearly, we must rely on the computational power of the computer.

It must be admitted, however, that Cohen's programs are like hack artists, who can draw only in a given style. The style may be rich enough (the generative system powerful enough) to make their drawings individually unpredictable. But the style itself is easily recognised. At present, only Cohen can change the constraints built into the program, so enabling it to draw pictures of a type that it could not have drawn before. But some programs, perhaps including some yet to be written by Cohen, might do so for themselves.

To be able to transform its style, a program would need (among other things) a metarepresentation of the lower-level constraints it uses. For the creative potential of a self-transforming system depends on how it represents its current skills (drawing 'a left arm and a right arm' or drawing 'two arms'), and on what heuristics are available to modify those representations and thereby enlarge its skills. We have already seen that if Cohen's program had an explicit representation of the fact that it normally draws four-limbed people, and if it were given very general 'transformation heuristics' (like 'drop a constraint', 'consider the negative' or 'vary the variable'), it might sometimes omit, or add, one or more limbs.

These remarks about creative potential apply to humans as well as to computer programs. Recent evidence from developmental psychology suggests that this sort of explicit representation of a lower-level drawing skill is required if a young child is to be able to draw a one-armed man or a seven-legged dog (Karmiloff-Smith 1990). Comparable evidence has been found with regard to other skills, such as language and piano playing; here too, imaginative flexibility requires the development of generative systems that explicitly represent lower-level systems (Clark and Karmiloff-Smith 1994; Karmiloff-Smith 1986). As for historical evidence, it is clear that the invention of new systems of representation, such as arabic numerals or musical notation, enormously increases the creative range of people using that representation.

MODELLING MUSICAL CREATIVITY

An example of an 'artistic' program grounded firmly in ideas about human psychology is the jazz improviser written by Philip Johnson-Laird (1988, 1993). This has appeared in no concert halls, and at first hearing seems much less impressive than Cohen's programs (Johnson-Laird likens its performance to that of 'a moderately competent beginner'). However, it raises some highly specific questions – and provides some suggestive answers about the nature of the complex conceptual space involved, and about how human minds are able to explore it.

A jazz musician starts with a chord sequence, such as a twelve-bar blues. (The performance will be an improvisation based on a fixed number of repetitions of the chord sequence.) Often, the chord sequence has already been written by someone else, for writing such sequences, unless they are kept boringly simple, typically requires a great deal of time and effort. They are complex hierarchical structures, with subsections 'nested' at several different levels, and with complex harmonic constraints linking sometimes far-separated chords. They could not be improvised 'on the fly' (where no backtracking is possible), but require careful thought and self-correction.

To take an analogy from language, consider this sentence: *The potato that the rat that the cat that the flea bit chased around the block on the first fine Tuesday in May nibbled is rotting.* You probably cannot understand this multiply nested sentence without pencilling in the phrase boundaries, or at least pointing to them. If someone were to read it aloud, without a very exaggerated intonation, it would be unintelligible. Moreover, you would find it difficult, perhaps impossible, to invent such a sentence without writing it down. For you cannot select the word *is* without remembering potato, 22 words before. (If you had started with *The potatoes*, you would have needed *are* instead.)

Similarly, jazz composers cannot improvise complicated chord sequences. Indeed, they have developed a special written notation to help them to keep the various harmonic constraints in mind while composing such sequences.

The jazz musician's task, in playing a chord sequence, is more difficult than yours in reading a sentence. For he is improvising, rather than merely reading. The 'chords' in the chord sequence are actually classes of chords, and the player must decide, as he goes along, just how to play each chord. He must also decide how to pass to the next chord, how to produce a melody, how to harmonise the melody with the chords, how to produce a bass-line accompaniment, and how to keep the melody in step with the metre.

Johnson-Laird argues that, because of the limited storage capacity of human short-term memory, the rules (or musical 'grammar') used for generating these features of the performance must be much less powerful than the hierarchical grammar used to produce chord sequences. Accordingly, his program consists of two parts.

One part generates a simple, harmonically sensible chord sequence (compare 'The potato is rotting'), and then complicates it in various ways to produce a nested hierarchical structure (comparable to a grammatically complex sentence). The second part takes that chord sequence as its input, and uses less powerful computational rules to improvise a performance in real time. What counts as an acceptable 'melody', for instance, is determined by very simple rules that consider only a few previous notes; and the harmonies are chosen by reference only to the immediately preceding chord.

When more than one choice is allowed by the rules, the program chooses at random. A human musician might do the same. Or he might choose according to some idiosyncratic preference for certain intervals or tones, thus giving his playing an 'individual' style. (The same obviously applies for literature and painting.) This is one of the ways in which chance, or randomness, can contribute to creativity. But it is the constraints – governing harmony, melody and tempo – that make the jazz performance possible in the first place. Without them, we would have a mere random cacophony.

Besides harmony, melody and tempo, there are other structures that inform music. Piano music, for example, is composed to be played expressively (composers often put expression marks in the score), and human musicians can play it with expression. Indeed, they have to: without expression, a piano composition sounds musically dead, even absurd. In rendering the notes in the score, pianists add such features as *legato*, *staccato*, *piano*, *forte*, *sforzando*, *crescendo*, *diminuendo*, *rallentando*, *accelerando*, *ritenuto* and *rubato* (not to mention the two pedals).

But how? Can we express this musical sensibility precisely? That is, can we specify the relevant conceptual space? Just what is a *crescendo*? What is a *rallentando*? And just how sudden is a *sforzando*?

These questions have been asked by Christopher Longuet-Higgins (whose earlier work on the conceptual space of tonal harmony was used within Johnson-Laird's jazz program [Longuet-Higgins 1987]). By means of a computational method, he has tried to specify the musical skills involved in playing expressively.

Working with Chopin's *Minute Waltz* and *Fantaisie Impromptu in C Sharp Minor*, Longuet-Higgins has discovered some counter-intuitive facts about the conceptual space concerned (Longuet-Higgins 1994). For example, a *crescendo* is not uniform, but exponential (a uniform *crescendo* does not sound like a *crescendo* all, but like someone turning up the volume knob on a radio); similarly, a *rallentando* must be exponentially graded (in relation to the number of bars in the relevant section) if it is to sound 'right'. Where *sforzandi* are concerned, the mind is highly sensitive: as little as a centisecond makes a difference between acceptable and clumsy performance. By contrast, our appreciation of *piano* and *forte* is less sensitive than one might expect, for (with respect to these two compositions, at least) only five levels of loudness are needed to produce an acceptable performance. More facts such as these, often demonstrable to a very high level of detail, have been discovered by Longuet-Higgins's computational experiments. As he points out, many interesting questions concern the extent to which they are relevant to a wide range of music, as opposed to a particular musical style.

Strictly speaking, this work is not a study of creativity. It is not even a study of the exploration of a conceptual space, never mind its transformation. But it is highly relevant to creativity (as is Longuet-Higgins's earlier computational work on harmony and musical perception [1987]). For we have seen that creativity can be ascribed to an idea only by reference to a particular generative system, or conceptual space. The more clearly we can identify this space, the more confidently we can identify and ask questions about the creativity involved in negotiating it. A pianist whose playing style sounds 'original', or even 'idiosyncratic', may be exploring and transforming the space of expressive skills that Longuet-Higgins has studied.

Of course, we can recognise this originality 'intuitively' and enjoy – or reject – the pianist's novel style accordingly. (Recognising it and describing it are two different things: the slow tempo of Rosalyn Tureck's performances of Bach is immediately obvious, but many other expressive characteristics of her playing are not.) Likewise, we can enjoy – or reject – drawings done by human artists or by computer programs. But understanding, in rigorous terms, *just how these creative activities are possible* is another matter. If that is our aim, computational concepts and computer modelling can help.

LITERARY SPACES

Literature involves many different conceptual spaces, mutually integrated in sensible – and sometimes surprising – ways. One of these concerns human motivation, the various psychological structures that are possible – and intelligible – within human action and interaction. Most novels and short stories are less concerned with transforming this space than with exploring it in a novel and illuminating fashion.

Current computer programs that write stories are woefully inadequate compared with human story tellers. But the best of them get what strength they possess from their internal models of very general aspects of motivation. Consider this example, written by a program asked to write a story with the moral 'Never trust flatterers':

> *The Fox and the Crow*
> Once upon a time, there was a dishonest fox named Henry who lived in a cave, and a vain and trusting crow named Joe who lived in an elm tree. Joe had gotten a piece of cheese and was holding it in his mouth. One day, Henry walked from his cave, across the meadow to the elm tree. He saw Joe Crow and the cheese and became hungry. He decided that he might get the cheese if Joe Crow spoke, so he told Joe that he liked his singing very much and wanted to hear him sing. Joe was very pleased with Henry and began to sing. The cheese fell out of his mouth, down to the ground. Henry picked up the cheese and told Joe Crow that he was stupid. Joe was angry, and didn't trust Henry any more. Henry returned to his cave. THE END.
> (Schank and Riesbeck 1981)

Exciting this little tale is not. But it does, as requested, show us that trusting flattery can lead to disappointment. The story has a clear structure and a satisfactory end. The characters have goals, and can set up subgoals to achieve them. They can cooperate in each other's plans, and trick each other so as to get what they want. They can recognise obstacles, and sometimes overcome them. They can ask, inform, reason, bargain, persuade and threaten. They can even adjust their personal relationships according to the treatment they get, rewarding rescue with loyalty or deception with mistrust. And there are no loose ends left dangling to frustrate us.

The reason is that this program can construct hierarchical plans, ascribing them to the individual characters according to the sorts of motivation (food preferences, for example) one would expect them to have. It can think up cooperative and competitive episodes, as it can give one character a role (either helpful or obstructive) in another's plan. These roles need not be allocated randomly, but can depend on background interpersonal relations (such as competition, dominance and familiarity). And it can represent different sorts of communication between the characters (such as asking or bargaining), which constrain what follows in different ways.

All these matters (like the body models in Cohen's line-drawing programs) are represented as abstract computational schemata. In addition, there are procedures and heuristics for integrating these schemata in sensible ways. The program as a whole is a generative system capable of producing a story structure, or plot, and of instantiating it in respect of specific incidents and characters.

A story writer equipped not only to do planning, but also to juggle with psychological schemata such as escape, ambition, embarrassment or betrayal could come up with better stories still. To design such a program would be no small feat. Every psychological concept involved in the plots of its stories, whether explicitly named in the text or not, would need to be defined – much as 'stability' had to be defined for the acrobat-drawing program, and 'melody' for the jazz improviser.

Ideally, these psychological concepts should allow for several different varieties, which could enter into story plots in significantly different ways. Consider betrayal, for instance, a concept that figures in many stories – from the court of the Moor of Venice to the Garden of Gethsemane. A very early computationally inspired definition of betrayal was: *Actor F, having apparently agreed to serve as E's agent for action A, is for some reason so negatively disposed toward that role that he undertakes instead to subvert the action, preventing E from attaining his purpose* (Abelson 1973). Suppose that the story writer had some representation of the facts that actors *in general* may vary in power, and that goals *in general* may vary in importance to one (specifiable) actor or another. The conceptual space of betrayal could then be explored by varying the importance (to one actor or the other) of the actions involved.

We can understand *abandonment* and *letting down*, for example, as distinct species of betrayal by 'tweaking' the definition given above. To accuse F of abandoning E is to say that he was acting initially as E's agent for action A (this action being crucial to E's welfare); that he has now deliberately stopped doing so; and that this amounts in effect, if not necessarily in intent, to the deliberate subversion of E's purposes – because E (by hypothesis) is helpless without F. In contrast, to say that F let E down implies neither the urgency of A nor the helplessness of E. In short, whereas anyone can let down or be let down, only the strong can abandon and only the weak can be abandoned. This is why abandonment is a peculiarly nasty form of betrayal.

Human authors, and readers, tacitly rely on such facts about the psychological structure of betrayal in writing and interpreting stories about it. They do likewise with respect to other psychological concepts. Some authors have the ability to make us recognise aspects of the relevant conceptual spaces that we had not seen before (much as Dickens reminds us that seven adjectives may accompany one noun). Henry James's novella *The Beast in the Jungle*, for example, is a superb depiction of a familiar motivational category instantiated in a subtly unfamiliar way. Not until the penultimate page does the reader realise just what the story has been about, just what psychological space it has been exploring. At that point, however, the matter becomes glaringly obvious (compare: '*Of course* a noun can have seven adjectives!').

These sorts of conceptual exploration could, in principle, be done by story-writing computer programs too. But the complexities are so great (and the background knowledge of the world so extensive) that it is unrealistic to expect there to be a computerised story writer that can perform at better than a hack level – if that. Our interest, however, is not in getting computers to do our creative acts for us, but in using the computational approach to help us understand what is involved when we do them.

ANALOGY

Analogy is widely employed in the arts. Literary analogies abound in prose and poetry, visual analogies enliven paintings, and kinetic analogies inform the ballet (think of the jerky actions of the doll Coppelia, or the feline movements of Puss-in-Boots).

Scientific thinking exploits analogy, too. In an earlier section, for example, we took for granted that Kekulé was capable of recognising the analogy between string molecules and 'long rows', and between twisting rows and snakes. Historians and philosophers of science have noted the importance of analogies in scientific discovery and theory (Hesse 1988). And Koestler (1964) held that the most creative moments in science involve the recognition of a novel analogy between previously unrelated fields.

How is analogical thinking possible? An analogy links two previously unrelated concepts. To understand how it arose, we must detail the psychological structure of the two concepts concerned, and specify processes whereby these two spaces can be simultaneously retrieved, compared and linked. Computational psychology offers some suggestions about what such processes might be like.

Relatively close analogies – family resemblances, Kekulé's rows-as-snakes, and much poetic imagery as well (Boden 1990: ch. 6) – may depend on processes broadly similar to those built into 'connectionist' computer systems. These parallel-processing systems, often called 'neural networks', are composed of many simple computational units, each coding one semantic feature (Rumelhart and McClelland 1986). The units are linked by excitatory and inhibitory connections (as are neurons in the brain). Units coding for mutually consistent features tend to excite each other's activity, whereas mutually inconsistent units inhibit each other. For instance, a unit coding for 'white' may excite both 'cream' and (less strongly) 'yellow', but it will inhibit 'blue', 'red' and (above all) 'black'.

Because of their basic design, connectionist systems can take many different constraints into account simultaneously, where no constraint is necessary but a large number are sufficient for making the judgement concerned. It follows that they are inherently tolerant of noise (missing or spurious information), and superior to traditional AI programs in their ability to associate similar but non-identical patterns. A connectionist system that has already learned the pattern *Mary had a little lamb*, for instance, will 'naturally' be able to retrieve that entire pattern if presented with the fragment *Mary had a*. . . . Likewise, it will 'spontaneously' recognise that *Martha had a little duck* is similar. In other words, the system can be reminded of a familiar idea by encountering a fragment of it, or by coming across a similar idea.

Those analogies in art and science that seem most creative, however, do not rely on reminding of this common type. They are more surprising, not to say highly counter-intuitive. Consider Macbeth's description of sleep:

> Sleep that knits up the ravelled sleeve of care,
> The death of each day's life, sore labour's bath,
> Balm of hurt minds, great nature's second course,
> Chief nourisher in life's feast.

This passage works because Shakespeare's readers, like him, know about such worldly things as knitting, night and day, and the soothing effects of a hot bath. In addition, they are able to understand analogies, even highly unusual or 'creative' analogies, such as comparing sleep with a knitter. But how can this be? A knitter is an animate agent, but sleep is not. How can the human mind map 'sleep' on to 'knitter' so as to realise the link: that both can repair the ravages of the previous day?

Similarly, how can we understand Socrates' remark (in Plato's *Theaetetus*) that the philosopher is 'a midwife of ideas'? A philosopher is not (usually!) a midwife. And while a new idea is indeed new, vulnerable and perhaps flawed – like a baby – it is nevertheless very different from a baby. Like sleep, ideas are not even animate. How, then, can someone create, or creatively interpret, such a strange comparison? Such a thought seems to be *impossible*.

Analogical thinking has been widely studied by psychologists, some of whom have produced computational models of it (Vosniadou and Ortony 1989). A number of connectionist systems have been specifically designed to interpret 'surprising' analogies, as opposed to mere family resemblances (Holyoak and Barnden, in press).

One such system was given structured representations of the concepts of philosopher and midwife, and was then presented with Socrates' analogy (Holyoak and Thagard 1989). It mapped 'idea' on to 'baby' as required. The model includes a large semantic network in which concepts are associated, as the concepts stored in human memory seem to be, somewhat in the way of a thesaurus. They bear links to synonyms, defining properties, and less closely related words such as opposites (so this network could support many different uses of 'consider the negative'). The analogy mapper compares concepts in terms of structural similarity, semantic centrality and pragmatic (contextual) importance. On being told that there is some (unspecified) analogy between 'philosopher' and 'midwife', this program mapped 'baby' on to 'idea' even though it recognised that a central feature of a baby (its being alive) does not hold of an idea.

This analogy interpreter has a 'sister system' that comes up with analogies, as opposed to interpreting ready-made analogies input to it (Thagard *et al.* 1988). It does come up with some fairly 'surprising' analogies (for instance, it notes the resemblance between the schematised plots of *Romeo and Juliet* and *West Side Story*). But in its current form, it would not spontaneously generate either the idea–baby or the sleep–knitter comparison, because it looks for the 'best' – that is, the closest – analogy it can find. Even if it were told to ignore the 20 best comparisons, it would not come up with either of these notions. Part of the reason is that its designers were most interested in analogy in science, where closeness is in general an advantage. In poetry, by contrast, *distance* between the two poles of the analogy is often preferred.

Even poetic distance, however, has to be kept within the bounds of intelligibility. Poets help us to interpret a far-distant analogy by providing additional constraints within the context of the poem. In the four-line fragment of Macbeth's speech, for instance, there is a succession of images for sleep each of which (even 'death') suggests some alleviation of previous troubles. The wildness of each individual analogy is thus tempered by the mutually reinforcing semantic associations set up by all the others.

Creative scientists, likewise, justify bold analogies by reference to the theoretical context concerned. Moreover, to accept a new scientific analogy is thenceforth to perceive the experimental situation in a new way. William Harvey's description of the heart as a pump changed not only what experiments were done, but how experimental events (such as systole and diastole) were perceived. The theory-laden nature of observation is a commonplace within the philosophy of science.

A psychology of analogy should be able to show how aptness to the current context can be achieved, and how a new analogy and a new perception can develop together. The analogy programs described above cannot help here, because their contextual sensitivity is shallow and their representations are fixed. After 'philosopher' has been mapped on to 'midwife', it is represented in exactly the same way as before; but Socrates' aim in introducing the analogy was not merely to point out a likeness, but to alter Theaetetus' perception of what a philosopher is. A computational model of analogy that focuses on these issues of context sensitivity and altered perception is Douglas Hofstadter's 'Copycat' (Chalmers *et al.* 1991; Hofstadter and Mitchell, in press; Hofstadter *et al.* 1987; Mitchell 1993).

Hofstadter stresses that one's perception of a situation is normally biased by high-level concepts and aims. Imagine three observers in the same room: the first may see the person in the corner as a woman holding wooden knitting needles, the second as a loving mother carefully mending her child's torn garment, and the third as a proletarian sweatshop worker exploited by the capitalist system. Indeed, these three observers may all be inside a single head: depending on one's interests at the time, one may see the scene in any of these ways. A fourth observer, currently writing a poem about overwhelming guilt, may focus on the steadily lengthening sleeve and be reminded of the refreshing powers of sleep. In each case, the representation of the situation is relevant to the beliefs and interests of the perceiver. Moreover, it is hard to say where perception ends and analogising begins.

The Copycat project takes these facts about human psychology seriously. The program allows for the generation of many different analogies, where contextually appropriate comparisons are favoured over inappropriate ones. It does not rely on ready-made, fixed representations, but constructs its own representations in a context-sensitive way: its new analogies and new perceptions develop together.

Copycat's 'perceptual' representations of the input patterns are built up dialectically, each step being influenced by (and also influencing) the type of analogical mapping that the current context seems to require. A partially built interpretation that seems to be mapping well on to the nascent analogy is maintained and developed further. A partially built representation that seems to be heading for a dead end is abandoned, and an alternative one started that exploits different aspects of the target concept. Varying degrees of conceptual 'slippage' are allowed, so that analogies of differing closeness can be generated.

The domain actually explored by Copycat is a highly idealised one, namely, alphabetic letter strings. But the computational principles involved are relevant to analogies in any domain. In other words, the alphabet is here being used as a psychological equivalent of inclined planes in physics.

Copycat considers letter strings such as *ppqqrrss*, which it can liken to strings such as *mmnnoopp*, *tttuuuvvvwww* and *abcd*. Its self-constructed 'perceptual' representations describe strings in terms of descriptors such as *leftmost*, *rightmost*, *middle*, *same*, *group*, *alphabetic successor* and *alphabetic predecessor*. It is a parallel-processing system, in that various types of descriptor compete simultaneously to build the overall description.

The system's sense of analogy in any particular case is expressed by its producing a pair of letter strings that it judges to be like some pair provided to it as input. In general, it is able to produce more than one analogy, each of which is justified by a different set of abstract descriptions of the letter strings.

For instance, Copycat may be told that the string *abc* changes into *abd*, and then asked what the string *mrrjjj* will change into. As its answer, it may produce any of the following strings: *mrrjjd*, *mrrddd*, *mrrkkk* or *mrrjjjj*. The last one is probably the one that you prefer, since it involves a greater level of insight (or abstraction) than the others. That is, it involves seeing *mrrjjj* as *m-rr-jjj*, and seeing the lengths of the letter groups, and then in addition seeing that the group lengths form a 'successor group' (1-2-3), and then finally seeing that '1-2-3' maps on to *abc*. At one level of abstraction, then, the analogy is this: *abc* goes to *abd*, and 123 goes to 124; but at the letter level (the level it was actually posed at), the analogy is this: *abc* goes to *abd*, and *mrrjjj* goes to *mrrjjjj*. But if this is the 'best' answer, the other answers are quite interesting. Is *mrrjjd* better than, worse than, or equivalent to *mrrddd*? Why is *mrrkkk* better than both of those? Why is *mrrjjjj* better than all of them? And why is *mrrkkkk* (with four letters *k*) *inferior* to *mrrjjjj*?

The mapping functions used by Copycat at a particular point in time depend on the representation that has already been built up. Looking for *successors* or for *repetitions*, for instance, will be differentially encouraged according to the current context. So the two letters *mm* in the string *ffmmtt* will be mapped as a sameness pair, whereas in the string *abcefgklmmno* they will be perceived as parts of two different successor triples: *klm* and *mno*.

Even in the highly idealised domain of alphabetic letter strings, interesting problems arise. Suppose, for instance, that Copycat is told that *abc* changes into *abd*, and it must now decide what *xyz* changes into. What will it say? (What would you say?)

Its initial description of the input pair, couched in terms of alphabetic successors, has to be destroyed when it comes across *z* – which has no successor. Different descriptors then compete to represent the input strings, and the final output depends partly on which descriptors are chosen. On different occasions, Copycat comes up with the answers *xyd*, *xyzz*, *xyy* and others. However, its deepest insight is when (on approximately one run out of eight) it chances to notice that at one end of one string it is dealing with the *first* letter of the alphabet, and at the other end of the other string, it is dealing with the *last*. This suddenly opens up a radically new way of mapping the strings on to each other: namely, with *a* mapping on to *z* and simultaneously *left* on to *right*. As a consequence of this conceptual reversal, *successor* and *predecessor* also swap roles, and so the idea of 'replacing the rightmost letter by its successor', which applied to the initial string, metamorphoses under this mapping into 'replace the *leftmost* letter by its *predecessor*'. This gives the surprising and elegant answer, *wyz*.

You will have noticed that the initial description in this case is not merely adapted, but destroyed. Hofstadter compares this example with conceptual revolutions in science: the initial interpretation is discarded, and a fundamentally different interpretation is substituted for it.

These ideas about the interdependence of analogy and perception can be informally applied to our previous example. A painter, looking at the knitting woman, might sense some analogy to the portrait of Whistler's mother. In building his perceptual representation, he might therefore concentrate in turn (guided by his memory of the portrait) on the living woman's bodily attitude, hairstyle and hair colour, and skirt length. A political activist would find nothing of interest in such matters. His representation of the scene might ignore the physical details entirely, focusing instead on the vulnerability, powerlessness and political ignorance of non-unionised female workers – going like lambs to the slaughter, as he might (analogically) say.

Neither of these observers would pick out the currently relevant aspects of the entire situation immediately, for neither (we assume) came to the scene with detailed foreknowledge of what he would find, still less of what analogical associations he would be wanting to make. Rather, they would pick out the relevant aspects continuously, by a dialectical process of interpretative-analogical thinking. Much, perhaps even all, of this context-sensitive construction would occur subconsciously. But conscious inference might play a role, especially if someone were puzzling to interpret an analogy as opposed to generating one spontaneously (maybe the politician heard the painter say 'Look! Whistler's mother!').

This constructive process can be 'telescoped'. Suppose that the painter and politician were told, before entering the room, that they would see something very like Whistler's mother. In that case, they would enter the room with certain mapping rules already prepared, and would see the expected analogy very quickly. Such telescoping enabled a positivist philosopher in the 1950s to play a practical joke on a group of 'ordinary language' philosophers. Positivists had been arguing for some years that when we look at a straight stick half-immersed in a glass of water, we *see* only 'sense data' (which include the appearance of a bend), and we then use our knowledge about refraction to *infer* that the sense data are caused by a straight stick, not a bent one. Their opponents had countered that there are no 'sense data', and that we can properly be said to *see*, and even to *know*, that the stick is straight. Predictably, when the positivist lecturer held up a glass of water with a stick in it, the linguistic philosophers in the audience looked at it and insisted that the stick was obviously straight. In fact, it was bent. (Copycat's processing can be telescoped too: if the relevant descriptors are marked beforehand, the system will use those descriptors in preference to others – even though it is still potentially capable of perceiving its data in many ways.)

Culturally based telescoping of this sort explains why a schoolchild can quickly understand, perhaps even discover, an analogy that took the relevant H-creative thinker many months or years to grasp. The particular analogy, we assume, is new to the child. But its general type is familiar. The notion that simple linear equations, for example, capture many properties of the physical world may already be well established in the pupil's mind. It is hardly surprising, then, if this analogical mapping mechanism can be activated at the drop of the teacher's chalk.

As Hofstadter points out, most current computational models of analogy (and of problem solving, including scientific discovery) put the computer in the place of the schoolchild. That is, the relevant representations and mapping rules are provided ready-made to the program. It is the programmer who has done the work of sifting and selecting the 'relevant' points from the profuse conceptual apparatus within his mind. Copycat, preliminary though it is, shows that a computational theory of creative thinking need not take relevance for granted in this way.

TRANSFORMATION IN MODELS OF SCIENTIFIC DISCOVERY

Computational work on scientific thinking is more common than work on artistic skills. Several 'inductive' programs have come up with useful (in some cases, H-novel) scientific ideas. For instance, a suite of programs designed to find simple mathematical and classificatory relations has 'rediscovered' many physical and chemical laws (Langley *et al.* 1987; Zytkow 1990). And an expert system (dealing with a strictly limited area of stereochemistry) has drawn chemists' attention to molecules they had not previously thought of (Lindsay *et al.* 1980). This system has even been listed in the acknowledgements of a refereed paper published in the *Journal of the American Chemistry Society* (Buchanan *et al.* 1976). Like most current AI systems (except Copycat), however, these 'discovery programs' depend on the programmers' prior handcrafting of the relevant data. What's more, like the systems discussed in the previous sections, they are exploratory rather than transformational.

Programs capable of *transforming* their own conceptual space are still few and far between. One such is the 'Automatic Mathematician' (AM) (Lenat 1983). This system does not produce proofs, or solve mathematical problems. Rather, it generates and explores mathematical ideas, coming up with new concepts and hypotheses to think about.

AM starts out with 100 very primitive mathematical concepts drawn from set theory (including sets, lists, equality and operations). These concepts are so basic that they do not even include the ideas of elementary arithmetic. To begin with, the program does not know what an integer is, still less addition, subtraction, multiplication and division.

Also, AM is provided with about 300 heuristics. These can examine, combine and transform AM's concepts – including any compound concepts built up by it. Some are very general, others specific to set theory, and they enable AM to explore the space potentially defined by the primitive concepts. This exploration involves conceptual change, by means of various combinations and transformations.

For example, AM can generate the *inverse* of a function. This heuristic (a mathematical version of 'consider the negative') enables the program to define multiplication having already defined division, or to define square roots having already defined squares. Another transformation generalises a concept by changing an 'and' into an 'or' (compare relaxing the membership rules of a club from 'anyone who plays bridge and canasta' to 'anyone who plays bridge or canasta').

However, AM does not consider *every* negative, nor change *every* 'and' into an 'or'. Time and memory do not allow this. Like all creative thinkers, AM needs hunches to guide it along some paths rather than others. And it must evaluate its hunches, if it is to appreciate its own creativity. Accordingly, some of AM's heuristics suggest which sorts of concept are likely to be the most interesting. If it decides that a concept is interesting, AM concentrates on exploring that concept. For example, it takes note if it finds that the union of two sets has a simply expressible property that is not possessed by either of them. This is a mathematical version of the familiar notion that *emergent* properties are interesting. In general, we are interested if the combination of two things has a property that neither constituent has.

AM's hunches, like human hunches, are sometimes wrong. Nevertheless, it has come up with some extremely powerful notions. It produced many arithmetical concepts, including *integer*, *prime*, *square root*, *addition* and *multiplication*. It generated, though of its nature could not attempt to prove, the fundamental theorem of number theory: that every number can be uniquely factorised into primes. And it suggested the interesting idea (Goldbach's conjecture) that every even number greater than two is the sum of two different primes.

It defined several concepts of number theory by following unusual paths – in two cases, inspiring human mathematicians to produce much shorter proofs than were previously known. It has even originated one minor theorem that no one had ever thought of before (concerning 'maximally divisible' numbers, which AM's programmer knew nothing about). In short, AM appears to be significantly P-creative, and slightly H-creative too.

Some critics have suggested that this appearance is deceptive, that some of the heuristics were specifically included to make certain mathematical discoveries possible. In reply, AM's programmer insists that the heuristics are fairly general ones, not special-purpose tricks. On average, he reports, each heuristic was used in making two dozen different discoveries, and each discovery involved two dozen heuristics. Even so, a given heuristic may have been used only once, in making an especially significant discovery. (A detailed trace of the actual running of the program would be needed to find this out.) The question would then arise whether it had been put in for that specific purpose, or for exploring mathematical space in a more general way. The precise extent of AM's creativity, then, is unclear. But we do have some specific ideas about what sorts of questions are relevant.

Whereas AM has heuristics for altering concepts, a successor program (EURISKO) possesses heuristics for changing heuristics. As a result, EURISKO can explore and transform not only its stock of concepts, but its own processing style.

For example, one heuristic asks whether a rule has ever led to any interesting result. If it has not (given that it has been used several times), it is marked as less valuable – which makes it less likely to be used in future. What if the rule has occasionally been helpful, though usually worthless? Another heuristic, on noticing this, suggests that the rule be specialised. The new heuristic will have a narrower range of application than the old one, so will be tried less often (thus saving effort). But it will be more likely to be useful in those cases where it is tried.

Moreover, the 'specialising heuristic' can be applied *to itself*. Because it is sometimes useful and sometimes not, EURISKO can consider specialising it in some way. The program distinguishes several sorts of specialisation, and has heuristics for all of them. Each is plausible, for each is often (though not always) helpful. And each is useful in many different domains. One form of specialisation requires that the rule being considered has been useful at least three times. Another demands that the rule has been *very* useful, at least once. Yet another insists that the newly specialised rule must be capable of producing all the past successes of the unspecialised rule. And a fourth heuristic specialises the rule by taking it to an extreme.

Other heuristics work not by specialising rules, but by generalising them. Generalisation, too, can take many forms. Still other heuristics can create new rules by analogy with old ones. Again, various types of analogy can be considered.

With the help of various packets of specialist heuristics to complement these general ones, EURISKO has been applied in several different areas. It has come up with some H-novel ideas, concerning genetic engineering and computer-chip (VLSI) design. Some of its ideas have even been granted a US patent (the US patent law insists that the new idea must not be 'obvious to a person skilled in the art').

The general principles of heuristic embodied within EURISKO have nothing specifically to do with science. They could, in theory, be applied to artistic spaces too. So some future version of the acrobat-drawing program, for example, might be able to alter – and even to transform – its graphic style by using methods like those described above. Like analogy, then, heuristic transformation is a general strategy of creative thinking that can be applied in many different fields of thought.

To be sure, scientific ideas have to fit the world in a way in which artistic ideas do not. But AM and EURISKO are not concerned with validation, or proof: their role is simply to come up with potentially interesting new ideas. The selection of the best ideas, and the weeding out of the worst, must be done by their human users.

GENETIC ALGORITHMS

The preceding remark should not be taken to mean that *only* humans, in principle, are able to sort the wheat from the chaff. But this is a very common view. It is expressed, for instance, in the following poem, fictionally ascribed by Laurence Lerner to an imaginary computer program called ARTHUR (Automatic Record Tabulator but Heuristically Unreliable Reasoner) (Lerner 1974):

Arthur's Anthology of English Poetry
To be or not to be, that is the question
To justify the ways of God to men
There was a time when meadow grove and stream
The dropping of the daylight in the west
Otters below and moorhens on the top
Had fallen in Lyonesse about their Lord.

There was a time when moorhens on the top
To justify the daylight in the west,
To be or not to be about their Lord
Had fallen in Lyonesse from God to men;
Otters below and meadow grove and stream
The dropping of the day, that is the question.

A time when Lyonesse and grove and stream
To be the daylight in the west on top
When meadow otters fallen about their Lord
To justify the moorhens is the question
Or not to be the dropping God to men
There was below the ways that is a time.

To be in Lyonesse, that is the question
To justify the otters, is the question
The dropping of the meadows, is the question

I do not know the answer to the question

There was a time when moorhens in the west
There was a time when daylight on the top
There was a time when God was not a question
There was a time when poets

Then I came

Lerner appears to believe that transformations that could be carried out by a computer program, such as 'mechanical' cut-and-paste, could not possibly generate anything sensible – and that no program could tell sense from nonsense anyway. The implication, so far as theoretical psychology is concerned, is that no computational theory could describe the generation of valuable new ideas, and that only an unanalysable faculty of 'intuition' or 'insight' could recognise their value. None of these beliefs is justified.

Consider, for example, a computer program that uses IF–THEN rules to regulate the transmission of oil through a pipeline in an economical way (Holland *et al.* 1986). It receives hourly measurements of the oil inflow, oil outflow, inlet pressure, outlet pressure, rate of pressure change, season, time of day, time of year and temperature. Using these data, it continually alters the inlet pressure to allow for variations in demand, infers the existence of accidental leaks, and adjusts the inflow accordingly.

So far, so boring. But – what is not boring at all – this program is a self-transforming system. It was not told which rules to use for adjusting inflow, or for detecting accidental leaks. It discovered them for itself. It started from a set of randomly generated rules, which it repeatedly transformed in part random, part systematic ways. To do this, it used heuristics called genetic algorithms. These enable a system to make changes that are both plausible and unexpected, for they produce novel recombinations of the most useful parts of existing rules.

As the name suggests, these heuristics are inspired by biological ideas about how the 'creative' process of evolution is effected. Some genetic changes are isolated mutations in single genes. But others involve entire chromosomes. For example, two chromosomes may swap their left-hand sides, or their midsections (the point at which they break is largely due to chance). If a chromosome contained only six genes, then the strings *ABCDEF* and *PQRSTU* might give *ABRSTU* and *PQCDEF*, or *ABRSEF* and *PQCDTU*. Such transformations can happen repeatedly, in successive generations. The strings that eventually result are unexpected combinations of genes drawn from many different sources. Genetic algorithms in computer programs produce novel structures by similar sorts of transformation.

Psychological applications of such simple combinatorial methods may seem doomed to failure. Indeed, these very methods are used by Lerner to ridicule the idea of a computer-poet. Almost all the lines in *Arthur's Anthology of English Poetry* are derived, by cut-and-paste recombinations, from the sixfold miscellany of the first verse. Starting with Shakespeare and Milton, the path runs steeply downward: the imaginary computer tells us that 'To justify the moorhens is the question', and produces the gnomic utterance, 'There was below the ways that is a time'.

Lerner's mockery of what are, in effect, genetic algorithms is not entirely fair, for many potentially useful structures were generated by his combinatorial method. Almost every line of his poem would be intelligible in some other verbal environment. 'To justify the moorhens is the question' might have occurred in *The Wind in the Willows*, if Ratty's friends had been accused of wrongdoing. Even 'Or not to be the dropping God to men' might have featured on Mount Olympus: 'Pick up your thunderbolt, Zeus! Do you want to be, or not to be, the 'dropping God' to men?' Only one line is utter gibberish: 'There was below the ways that is a time.'

The explanation is that Lerner swapped grammatically coherent fragments, rather than single words. A similar strategy was followed by those eighteenth-century composers (including Mozart) who wrote 'dice music', in which a dozen different choices might be provided for every bar (as opposed to every note) of a sixteen-bar piece. In general the plausibility of the new structures produced by this sort of exploratory transformation is increased if the swapped sections are coherent minisequences.

However, there is a catch - or rather, several. The first is that a self-adapting system must somehow identify the most useful 'coherent minisequences'. But these never function in isolation: both genes and ideas express their influence by acting in concert with many others. The second is that coherent minisequences are not always *sequences*. Co-adapted genes (which code for biologically related functions) tend to occur on the same chromosome, but they may be scattered over various points within it.

Similarly, potentially related ideas are not always located close to each other in conceptual space. Finally, a single unit may enter more than one group: a gene can be part of different co-adaptive groups, and an idea may be relevant to several kinds of problem.

Programs based on genetic algorithms help to explain how plausible combinations of far-distant units can nevertheless happen. They can identify the useful parts of individual rules, even though these parts never exist in isolation. They can identify the significant interactions between rule parts (their mutual coherence), even though the number of possible combinations is astronomical. And they can do this despite the fact that a given part may occur within several rules. Their initial IF–THEN rules are randomly generated (from task-relevant units, such as *pressure*, *increase* and *inflow*), but they can end up with self-adapted rules rivalling the expertise of human beings.

The role of natural selection is modelled by assigning a 'strength' to each rule, which is continually adjusted according to its success (in controlling the pipeline, for instance). The relevant heuristic is able, over time, to identify the most useful rules, even though they act in concert with many others – including some that are useless, or even counterproductive. The strength measure enables the rules to compete, the weak ones gradually dropping out of the system. As the average rule strength rises, the system becomes better adapted to the task environment.

The role of variation is modelled by heuristics (genetic operators) that transform the rules by swapping and inserting parts in ways like those outlined above. For instance, the 'crossover' operator swaps a randomly selected segment between each of two rules. Each segment may initially be in a rule's IF section or its THEN section. In other words, the crossover heuristic can change either the conditions that result in a certain action, or the action to be taken in certain conditions, or both.

One promising strategy would be to combine the effective components of several high-strength rules. Accordingly, the genetic operators pick only rules of relatively high strength. But the effective components must be identified (a rule may include several conditions in its IF side and several actions in its THEN side). The program regards a component as effective if it occurs in a large number of successful rules. A 'component' need not be a sequence of juxtaposed units. It may be, for instance, two sets of three (specified) neighbouring units, separated by an indefinite number of unspecified units. The huge number of possible combinations do not have to be separately defined, or considered in strict sequence. In effect, the system considers them all in parallel (taking into account its estimate of various probabilities in the environment concerned).

Contrary to Lerner's rhetorical intention, *Arthur's Anthology* shows that simple recombinations of ideas and conceptual themes can sometimes lead to potentially valuable ideas. To that extent, a combination theory may help to explain some examples of creative thinking. But, as remarked in the first section, a combination theory should show how these combinations can come about, and how the results can be selectively sifted. Work on genetic algorithms suggests that unconscious, non-deliberative psychological processes might enable largely random (but useful) combinations and sensible selections to be made in human minds.

Some visual artists are using evolutionary programs to help them produce images that – they assure us – they could not have imagined otherwise. Karl Sims's computer graphics program, for instance, uses genetic algorithms to generate new images from pre-existing images (Sims 1991; see also Todd and Latham 1992). In this case, the selection of the 'fittest' examples is not automatic. Instead, the programmer selects the images that are aesthetically pleasing, or otherwise interesting, and these are used to 'breed' the next generation. (Sims could provide automatic selection rules, but has not yet done so – not only because of the difficulty of defining aesthetic criteria, but also because he aims to provide an interactive graphics environment in which human and computer can cooperate in generating otherwise inconceivable images.)

In a typical run of the program, the first image is generated at random (but Sims can feed in a real image, such as a picture of a face, if he wishes). Then the program makes 19 independent changes (mutations) in the initial image-generating rule, so as to cover the VDU-screen with 20 images: the first, plus its 19 'asexually' reproduced) offspring. At this point, the human uses the computer mouse to choose either *one* image to be mutated, or *two* images to be 'mated' (through crossover). The result is another screenful of 20 images, of which all but one (or two) are newly generated by random mutations or crossovers. The process is then repeated, for as many generations as one wants.

How does Sims's program manage to tweak and transform image space? It starts with a list of 20 very simple LISP functions. A 'function' is not an actual instruction, but an instruction schema: more like '$x + y$' than '$2 + 3$'. Some of these functions can alter parameters in pre-existing functions: for example, they can divide or multiply numbers, transform vectors, or define the sines or cosines of angles. Some can combine two pre-existing functions, or nest one function inside another; so multiply nested hierarchies (many-levelled spaces) can eventually result. A few are basic image-generating functions ('maps' or images), capable, for example, of generating an image consisting of vertical stripes. Others can process a pre-existing image, for instance by altering the light contrasts so as to make 'lines' or 'surface edges' more or less visible.

Significantly, one may not be able to say just why *this* image resulted from *that* LISP expression. Sims himself cannot always explain the changes he sees appearing on the screen before him, even though he can access the miniprogram responsible for any image he cares to investigate, and for its parent(s) too. Often he cannot even 'genetically engineer' the underlying LISP expression so as to get a particular visual effect. This is partly because his system makes several changes simultaneously, with every new generation.

Where human minds are concerned, we may similarly have multiple interacting changes (and no program explanation at our fingertips). These multiple changes and simultaneous influences arise from the plethora of ideas within the mind. Think of the many different thoughts that arise in your consciousness, more or less fleetingly, when you face a difficult choice or moral dilemma. Consider the likelihood that many more conceptual associations are being activated unconsciously in your memory, influencing your conscious musings accordingly. Even if we had a listing of all these influences, we might be in much the same position as Sims, staring in wonder at one of his nth-generation images and unable to say why *this* LISP expression gave rise to it. In fact, we cannot hope to know about more than a fraction of the ideas aroused in human minds (one's own, or someone else's) when such choices are faced. The notorious unpredictability, and even *post hoc* inexplicability, of human creativity would therefore be expected, if processes like genetic algorithms are going on in the mind.

This is not to say that the variational/combinatorial processes in human minds are closely similar to those in the pipeline program or Sims's computer graphics system. Like the other programs discussed earlier, these two examples are crude at best and mistaken at worst, when compared with human thinking. But current computational models do offer us some promising, and precise, ideas about how to identify, map, explore and transform conceptual spaces. And that, I have argued, is what the psychology of creativity is all about.

CAN CREATIVITY BE MEASURED?

Assuming that creativity can be identified, and even explained, can it also be measured? The basic meaning of the term applies to ideas. People and social groups are called creative only if they are thought to have produced creative ideas. If we could measure creative ideas, we could develop some way of 'counting' them so as to measure the creativity of individuals or cultures.

Our question, then, is whether – and if so, how – we can say that one creative idea is more, even much more, creative than another. To put the question in a more paradoxical way, but one that seems justified by the account of creativity given above, how can one impossible idea be more impossible than another?

One common usage of 'more creative' can be discarded, for present purposes, immediately. We saw in the first section that *creative* may be used as an honorific label reserved for H-creativity, as opposed to P-creativity. In that case, any H-creative idea is 'more creative' than any merely P-creative idea. Indeed, the latter would not be regarded as creative at all.

This restrictive sense of the term, applicable only to first-time historical novelty, is unhelpful here. Quite apart from the ubiquitous problem of the reliability of historical evidence, the point at issue here is not 'Who thought of X first?' but 'Is X a creative idea, and if so, how creative is it?' Our concern (as in previous sections) must be with P-creativity in general, of which H-creativity is a special case. If one wants to measure H-creativity (in comparing cultures, for instance), one must first find a way of measuring P-creativity and then apply it selectively to H-novel cases.

If by 'measurement' is meant the application of a numerical scale, based on one or a few numerically describable dimensions, then my account of creativity implies that the creativity of an idea cannot be measured. One cannot capture the interesting differences between the *Mona Lisa* and the *Demoiselles d'Avignon*, for instance, by a set of measurements. Certainly, the spatial area of Nicholas Hilliard's miniatures, or the light reflectance of Rembrandt's portraits, might be relevant to judgements about the originality of those two artists. But the most significant aesthetic questions about their paintings concern other features, grounded in structural properties of various kinds. The same is true of originality in science. In general, one cannot assess creative ideas by a scalar metric.

Some form of complexity measurement, as used by computer science, would be useful. However, depth within the space must be recognised too. The appropriate method of assessment would have to take into account the fact that conceptual spaces are multidimensional structures, where some features are 'deeper', more influential, than others. The prairie house fireplace, for example, is architecturally deeper than the bedroom, and much deeper than the (merely ornamental) balcony. Analogously, the linguist's *NP* is syntactically deeper than *determiner*, and much deeper than *red*. And the home key is harmonically deeper than a modulation from major to minor, and much deeper than a plagal cadence.

Moreover, daring harmony can coexist with conservative melody: how is the one to be weighed against the other? What about novels and poems: does the originality lie in the plot, the theme, the language, the meter, the imagery, the psychological insight, the political awareness . . . all or none of these, and/or something else? To compare the degree of creativity of two ideas, we would have to weigh depth against number: novelty in one deep feature (a core dimension of the space) might outweigh several simultaneous novelties in more superficial features.

Creative transformations would have to be compared in respect of their depth, and distinguished from mere superficial tweaking. This could best be done for ideas within a single domain, where the conceptual space is shared. But chalk could sometimes be compared with cheese: to put seven fireplaces in a prairie house is clearly more daring (it results in more significant structural differences) than to put seven adjectives with one noun, or to superimpose seven decorative trills on a melody.

This is not to dismiss the more superficial aspects of our thinking as evaluatively irrelevant. Balconies can be not only well placed (in relation to the overall structure), but beautifully wrought. Dickens's seven adjectives for the sinner Scrooge were well chosen (and well ordered). And baroque music delights us with its profuse ornamentation. Indeed, these 'superficialities' have their own internal structural principles. The wrought-iron balcony can be aesthetically evaluated as an object in itself, quite apart from its relation to its parent building. Musical ornamentation has its own structure, quite apart from its relation to the melody: our delight at Alfred Deller's singing, for instance, is elicited partly by his daring – and teasing – mordants, appoggiaturas and trills.

Comparative assessments of creativity must recognise that many creative achievements involve exploration, and perhaps tweaking, of a conceptual space, rather than radical transformation of it. The more complex the space, the greater its exploratory potential, the more 'mere' exploration will be valued. (This is true only up to a point: if the space is so complex as to be unintelligible to us, even in an intuitive way, its generative products will be rejected. The common phenomenon of initial scorn followed by universal acceptance reflects the difficulty people sometimes have of relating a new idea to the underlying space that generated it.)

The exploratory activities of normal science, for instance, are not uncreative, even though they do not involve the fundamental perceptual reinterpretations typical of scientific revolutions. Nobel Prizes are awarded not for revolutionary work in the Kuhnian sense, but for ingenious and imaginative problem solving that may involve fairly deep theoretical transformations (of string molecule to ring molecule, for instance). To call this scientific work mere puzzle solving is to risk losing sight of the distinction between following a well-marked path for three (or 300) more steps, and carving out a new path within territory that has been mapped only on a large scale.

It is significant, here, that some musicians regard Mozart as a greater composer than Haydn *even though* they allow that Haydn was more adventurous, more ready to transform contemporary musical styles. Mozart's superiority, on this view, lay in his fuller exploration (and tweaking) of musical space, his ability to amaze us by showing us what unsuspected glories lie within this familiar space. Whether this musical judgement is faithful to Haydn's and Mozart's work is irrelevant. The point is that it is one that can intelligibly be made. It follows that no creativity metric could be adequate that ignored structural exploration, focusing only on structural transformation.

Computational models of concepts within connectionist semantic networks sometimes provide a basis for a metric of conceptual distance. But 'metric' is perhaps a misleading term, because this is a structured distance. Copycat's measurements of analogical similarity between letter strings, for instance, take note of various sorts of structural likeness and dissimilarity. It is therefore able, as presumably you are too, to see that the alphabet-reversing *wyz* is a more creative response than is *xyd* to the input problem *abc* → *abd*; *xyz* → ???. Two concepts may be compared, for example, in terms of their abstractly defined internal structure and/or their specific semantic content and/or their customary associations. These comparisons may make it possible to compare two theories said to be incommensurable by Kuhnians. (A co-author of the program presented with the philosopher–midwife analogy has argued that a computational definition of 'explanatory coherence' can show how the 'incommensurable' theories of phlogiston and oxygen can be rationally compared [Thagard 1989].)

Lacking any explicit account of the relevant conceptual spaces, someone may nevertheless make intuitive judgements about creativity. ('I don't know anything about art, but I know what I like!') Some of those judgements may be well grounded, and the more experience the person has of the relevant genre, the more sensitive they are likely to be. But even the well-grounded judgements will be largely indefensible, in the sense that the person is unable to defend them in terms of explicitly identified features of the conceptual space concerned. The intuitions of someone with access to verbal descriptions of the nature and history of the genre will, in general, be more discriminating (so the study of art history can increase one's appreciation of art, not merely one's knowledge about it).

For the purpose of comparing ('measuring') creativity, however, verbal descriptions may not be enough. The more explicit we can be in describing the creativity concerned, the better.

In a few cases, computational analyses exist that make clear the depth and mutual influences of different parts or dimensions of the relevant conceptual space. The creativeness of using seven prairie fireplaces may have been sensed by architectural historians, able only to remark on the 'occult' properties of spatial balance involved. But now it can be explicitly described and explained in terms of the architectural grammar of prairie houses. Similarly, a musical grammar of jazz can show which chord sequences are structurally more complex than others (and how), and which improvisations relate to which aspects of the chord sequence. And a model of inductive reasoning within stereochemistry may show, more clearly than the pre-existing chemical theory, how different molecular structures are related to each other.

Such analyses draw on conventional (non-computational) work in aesthetics, musicology, and the history and philosophy of art and science – all of which aim to uncover the styles, genres and theoretical forms of human achievement. This is hardly surprising, for only an expert in a given domain can write interesting programs modelling that domain. It is no accident that Cohen is an acclaimed painter in his own right, that Johnson-Laird is a good amateur jazz pianist, or that the designers of the stereochemistry and 'explanatory coherence' programs include philosophers of science. Traditional work in the humanities is highly relevant to the computational understanding of creativity.

In sum: The computational approach to creativity is grounded in the more familiar disciplines. But it has a higher standard of explicitness and rigour, and a fiercer discipline of theory testing. The price it pays for these theoretical goods is limitation. As yet there are very few computational models of interesting conceptual spaces, and still fewer of creative transformations. Today's computational psychology is therefore of limited use in comparing the creativity of different creative products. Its contribution is to help point the way to the sorts of comparisons that we should be making.

REFERENCES

Abelson, R.P. (1973) 'The structure of belief systems.' In R.C. Schank and K.M. Colby (eds) *Computer Models of Thought and Language*, pp. 287–340. San Francisco: Freeman.

Boden, M.A. (1990) *The Creative Mind: Myths and Mechanisms*. London: Weidenfeld & Nicolson.

Brannigan, A. (1981) *The Social Basis of Scientific Discoveries*. Cambridge: Cambridge University Press.

Buchanan, B.G., Smith, D.H., White, W.C., Gritter, R., Feigenbaum, E.A., Lederberg, J. and Djerassi, C. (1976) 'Applications of artificial intelligence for chemical inference. XXII automatic rule formation in mass spectrometry by means of the meta-dendral program.' *Journal of American Chemistry Society* 98: 6168–6178.

Carroll, L. (1859) *Alice's Adventures in Wonderland*. London: Macmillan.

Chalmers, D.J., French, R.M. and Hofstadter, D.R. (1991) *High-Level Perception, Representation and Analogy: A Critique of Artificial Intelligence Methodology*. CRCC Technical Report 49. Center for Research on Concepts and Cognition, Indiana University, Bloomington, Indiana.

Clark, A. and Karmiloff-Smith, A. (1994) 'The cognizer's innards.' *Mind and Language* 8: 487–519.

Findlay, A. (1965) *A Hundred Years of Chemistry* (3rd edn) ed. T.I. Williams. London: Duckworth.

Hesse, M.B. (1988) 'Theories, family resemblances, and analogy.' In D.H. Helman (ed.) *Analogical Reasoning*, pp. 317–340. Amsterdam: Kluwer.

Hofstadter, D.R. and Mitchell, M. (in press) 'An overview of the Copycat project.' In K.J. Holyoak and J.A. Barnden (eds) *Advances in Connectionist and Neural Computation Theory, Vol. 2: Analogical Connections*. Norwood, NJ: Ablex.

Hofstadter, D.R., Mitchell, M. and French, R.M. (1987) *Fluid Concepts and Creative Analogies: A Theory and its Computer Implementation*. CRCC Technical Report 18. Center for Research on Concepts and Cognition, Indiana University, Bloomington, Indiana.

Holland, J.H., Holyoak, K.J., Nisbett, R.E. and Thagard, P.R. (1986) *Induction: Processes of Inference, Learning, and Discovery*. Cambridge, MA: MIT Press.

Holyoak, K.J. and Barnden, J.A. (eds) (in press) *Advances in Connectionist and Neural Computation Theory, Vol. 2: Analogical Connections*. Norwood, NJ: Ablex.

Holyoak, K.J. and Thagard, P. (1989) 'Analogical mapping by constraint satisfaction.' *Cognitive Science* 13: 295–356.
Johnson-Laird, P.N. (1988) *The Computer and the Mind: An Introduction to Cognitive Science.* London: Fontana.
Johnson-Laird, P.N. (1993) 'Jazz improvisation: a theory at the computational level.' In P. Howell, R. West and I. Cross (eds) *Representing Musical Structure*, pp. 271–326. London: Academic Press.
Karmiloff-Smith, A. (1986) 'From meta-processes to conscious access: evidence from children's metalinguistic and repair data.' *Cognition* 23: 95–147.
Karmiloff-Smith, A. (1990) 'Constraints on representational change: evidence from children's drawing.' *Cognition* 34: 57–83.
Koestler, A. (1964) *The Act of Creation.* London: Hutchinson.
Koning, H. and Eizenberg, J. (1981) 'The language of the prairie: Frank Lloyd Wright's prairie houses.' *Environment and Planning B* 8: 295–323.
Langley, P., Simon, H.A., Bradshaw, G.L. and Zytkow, J.M. (1987) *Scientific Discovery: Computational Explorations of the Creative Process.* Cambridge, MA: MIT Press.
Lenat, D.B. (1983) 'The role of heuristics in learning by discovery: three case studies.' In R.S. Michalski, J.G. Carbonell and T.M. Mitchell (eds) *Machine Learning: An Artificial Intelligence Approach.* Palo Alto, CA: Tioga.
Lerner, L.A. (1974) *A.R.T.H.U.R.: The Life and Opinions of a Digital Computer.* Hassocks, Sussex: Harvester Press.
Lindsay, R., Buchanan, B.G., Feigenbaum, E.A. and Lederberg, J. (1980) *DENDRAL.* New York: McGraw-Hill.
Longuet-Higgins, H.C. (1987) *Mental Processes: Studies in Cognitive Science.* Cambridge, MA: MIT Press.
Longuet-Higgins, H.C. (1994) 'The structural determinations of musical expression.' In M.A. Boden and A. Bundy (eds) *Artificial Intelligence and the Mind: New Breakthroughs or Dead-Ends?*, pp. 103–113. London: The Royal Society and The British Academy.
McCorduck, P. (1991) *Aaron's Code.* San Francisco: W.H. Freeman.
Mitchell, M. (1993) *Analogy as Perception.* Cambridge, MA: MIT Press.
Rumelhart, D.E. and McClelland, J.L. (eds) (1986) *Parallel Distributed Processing: Explorations in the Microstructure of Cognition.* 2 vols. Cambridge, MA: MIT Press.
Schank, R.C. and Riesbeck, C.K. (eds) (1981) *Inside Computer Understanding: Five Programs Plus Miniatures.* Hillsdale, NJ: Erlbaum Press.
Sims, K. (1991) 'Artificial evolution for computer graphics.' *Computer Graphics* 25 (4), July: 319–328.
Stiny, G. (1977) 'Ice-ray: a note on the generation of Chinese lattice designs.' *Environment and Planning B* 4: 89–98.
Stiny, G. (1991) 'The algebras of design.' *Research in Engineering Design* 2: 171–181.
Stiny, G. and Gips, J. (1978) *Algorithmic Aesthetics: Computer Models for Criticism and Design in the Arts.* Berkeley: University of California Press.
Stiny, G. and Mitchell, W.J. (1978) 'The Palladian grammar.' *Environment and Planning B* 5: 5–18.
Stiny, G. and Mitchell, W.J. (1980) 'The grammar of paradise: on the generation of Mughul gardens.' *Environment and Planning B* 7: 209–226.

Thagard, P. (1989) 'Explanatory coherence.' *Behavioral and Brain Sciences* 12: 435–502.

Thagard, P., Holyoak, K.J., Nelson, G. and Gochfeld, D. (1988) 'Analog retrieval by constraint satisfaction.' Research paper, Cognitive Science Laboratory, Princeton University.

Todd, S. and Latham, W. (1992) *Evolution Art and Computers*. London: Academic Press.

Vosniadou, S. and Ortony, A. (eds) (1989) *Similarity and Analogical Reasoning*. Cambridge: Cambridge University Press.

Zytkow, J.M. (1990) 'Deriving laws through analysis of processes and equations.' In P. Langley and J. Shrager (eds) *Computational Models of Discovery and Theory Formation*, pp. 129–156. San Mateo, CA: Morgan Kaufmann.

CHAPTER FOUR

CREATIVE THOUGHT

A long-term perspective

IAN HODDER

The concept of creativity might at first seem nebulous. In making an attempt to define what is meant by the term, it might help to draw comparisons with other akin words such as intelligence, adaptation and agency. One definition of intelligence might be 'the ability to adapt'. Human adaptation can perhaps be said to involve the creative use of information in novel circumstances. Human intelligence and adaptability, involving the ability to identify problems, select relevant information and find solutions, are closely bound up with creativity.

If creativity lies at the heart of what it means to be human, how does it differ from other foundational characteristics such as intelligence and adaptation? Presumably the answer lies in the emphasis on novelty. Often it seems that the term creativity is used synonymously with change. We could say that creativity arises out of the tension between imagination and rules. But one of the recurrent themes in this chapter will be whether the ability imaginatively to create novelty and change should be separated from adaptation and intelligence.

What distinguishes creativity from the intelligent adaptation of individuals in the daily practices of their lives? We can perhaps turn to another akin word and concept. The term 'agency' has assumed an importance in much contemporary social theory, including archaeology (e.g. Barrett 1994; Shanks and Tilley 1987). Various definitions might be given, but an important aspect of most definitions would be the ability to mobilise resources in the practices of daily life. There are perhaps two aspects of this definition that link to creativity.

First, agency relates to the notion of action. The latter can be distinguished from behaviour by the emphasis on intentionality. Thus 'I raise my hand' involves intentional action, whereas 'my hand is raised' describes behaviour. Intentional action is forward looking and goal oriented. It implies an evaluation of the consequences of behaviour such that choices can be made as to appropriate action in relation to some goal. As such it involves problem solving as solutions are sought regarding appropriate behaviour. Finding the right thing to do or the right thing to say clearly involves creativity, since all action takes place in changing contexts within a space–time continuum. The contingent aspects of all social life imply that creativity is closely linked to agency. Intentions have to be weighed against contingent context and creative solutions found. Second, however, indissolubly linked to intentionality and problem solving are the creative processes of interpretation and understanding. Determining appropriate action in a given context suggests the ability to evaluate contexts, to define appropriate action and to assess consequences. So agency involves interpreting, understanding, making sense of what is said and done in the world around us. It depends on the listening ear, or the steady eye, watching, intelligently making sense of things. This is not a passive process. It involves creatively linking things together, discerning intentions, observing trends, abstracting, conceptualising – indeed a whole complex of creative skills that are grouped together as 'interpretation'.

Creativity might be a creative way to think about aspects of the social process but I doubt that there is much substance to something called 'creativity' beyond intelligence, adaptability, agency, interpretation and problem solving. It would perhaps be better to say that there are various types of creativity. As I will argue below, what we mean by creativity changes historically (as for example in the definition of 'artistic creativity') as may well the substance of the various creative processes.

At the most general level, we can suggest that creativity is an essential component in all aspects of human existence because of its role in two areas. The first is related to intentionality and is proactive, forward-looking and problem solving. The second is related to perception, interpretation and problem recognition. At a still more general level, one might argue that the first is part of speech and action and the second is part of listening and watching. I take these two processes to be separable in terms of creative intelligence because of the manifest difference between the ability to recognise and make sense of the world around us and the ability to act and speak. Thus one might often recognise a face, or a word, a strain of music, a sound or a smell but be unable to articulate that knowledge in speech and action. People can usually recognise the meanings of many more words than they use in daily speech. Recognition is often easier than recall. I shall return to this point below.

CREATIVITY AS A SOCIAL PROCESS

I want to stress that creativity (perhaps of different types) is central to the two types of process identified above. Creativity is most obviously and most commonly associated with the more active processes of problem solving, imagination and invention. But I want to emphasise that the solutions to problems that are deemed most creative are usually those that fit *best* into existing schemes. We are often most impressed by solutions that most neatly transform our existing understanding. Indeed much creativity is about making links between bits of information rather than creating new bits or nodes. The need for solutions to resonate with contemporary concerns suggests that active problem solving is intimately linked to the process of interpretive understanding.

This emphasis on resonance and the overall links between creativity, agency and interpretation demonstrate that the creative process can only minimally be represented as cognitive. If we reduce creativity simply to the cognitive, we are left with very little. The problem can readily be appreciated by considering the perception of creativity or novelty. Notions of 'the creative spirit' often conjure up images of the artist. Whether we emphasise creativity in the arts or not, the identification of creativity is undoubtedly subjective and historical. It is often only by looking backwards that we can see that certain steps had sufficient impact to be called 'creative' or novel. Hobsbawm and Ranger (1983) have described the 'invention of tradition'. But it is equally true that creativity is invented by a backward looking gaze in order to make sense of or legitimate events in the present. The 'rediscovery' of classical art in the Renaissance or of Celtic art and its 'creativity' in the Romanticism of the nineteenth century (Merriman 1987) provide examples.

So creativity is not simply cognitive. Two artists may produce equally novel work but one of them might be seen as more creative than the other because the work resonates more fully with contemporary concerns and issues. This idea of resonance is central to our modern ideas of creativity and it underlines the social construction of creativity: it is indeed a meaningless concept unless it is related to social context. As another example, Lemonnier (1989) has described attempts to develop new forms of aeroplane. A wide range of technological innovations have been identified and were tried. But few have been accepted and incorporated into the modern aeroplane. Lemonnier argues that at least part of the answer for the limited creativity that is discernible in this field is that people want to fly in something that looks like an aeroplane. In other words, perceptions of what an aeroplane looks like limit the degree to which creativity is possible. Similar points concerning the social embeddedness of technological and scientific innovation have been made more widely (Latour and Woolgar 1979; Lemonnier 1993; van der Leeuw 1989).

If creativity is socially constructed and is partly at least in the eye of the beholder, how can we select out certain aspects of the social process and define them as creative? Clearly a purely cognitive approach runs into difficulties here. I am suspicious of Boden's (1990) distinction between impossibilist and associational creativity because these two types depend on an evaluation of what constitutes the transformation of a conceptual space. Any such evaluation will depend on social and strategic positioning in relation to the creative claims being made. As Hesse (1995) has shown, scientific advancement is heavily dependent on analogical or associational processes of thought. Claims for novelty depend on networks of relations within the scientific community (Latour and Woolgar 1979) and on historical evaluation. I would also point out that the distinction between P-creativity and H-creativity appears to depend on historical evaluation – that is on whether a new idea has wide historical implications. The social embeddedness of creativity is thus reinforced. The same criticism can be made of cognitive archaeology as a whole (Hodder 1993). This reduced and cognitive view of mind needs to be extended by a critique of the mind–body duality (Turner 1984) and with an integration of mind into an adequate social theory. I want to explore the social nature of creativity by considering what archaeology might contribute to such a theory, with particular reference to the later phases of prehistory. I see any such contributions linked to the two key distinctive aspects of archaeological enquiry – the long term and the focus on material culture.

CREATIVITY OVER THE LONG TERM

The similarity between the earliest automobiles and horse-drawn carriages is remarkable. The driver often sits in the back with the driving wheel in the centre of the car. The overall design and the arrangement and placing of the people travelling mirror the design and arrangement in a horse-drawn vehicle. The earliest clothes washing machines and the earliest vacuum cleaners look like the pre-motorised hand-driven versions. Major technological change often takes place gradually. The same gradualism is found with stylistic features. Since the late nineteenth century, Coca Cola bottles have changed only gradually, and in the first place the shape of the moulded bottle was derived from the endocarp (seed husk) of the coca plant. Innovation is often or always created through analogy and metaphor – through associations of ideas, so that change is slow and gradual. Much the same can be said of changes in Cadillac car wings, gradually transforming the metaphor of the rocket (Figure 4.1).

As a result, in the archaeological record, skeuomorphism is everywhere. It is argued that clay pots mimic skin or wooden containers, that corded decoration on pots derives from the cords of the containers in which pots were carried, and that flint daggers copy metal versions. Lug or strap handles on pots become residual decorative versions. Indeed, the whole of the typological method as used in archaeology could be said to be based on the assumption that things change only gradually. The 'battle-ship curve' describes the gradual increase in popularity and then decline of particular traits. Seriation is based on this assumption – that assemblages closer in time or space will be more similar than those placed farther apart. The underlying view is that change often occurs by a sort of cultural 'drift', and that the force of tradition is high.

There are many specific examples of routine ways of doing things that continue over very long periods of time. For example, in Mesoamerica, Flannery and Marcus (1983) argue that ethnographic organisation of settlement space into four colour-coded corners can be identified in the archaeological record perhaps back to Formative times. For the Andes, Lechtmann (1984) has identified a social and ritual role for metallurgy that contrasts with the wider use of metal tools in Europe. For large parts of the Americas, a particular complex of ideas connecting smoke and birds has been argued by von Gernet and Timmins (1987) to extend over enormous expanses of space and time. In southern Africa, Huffman (1984) has suggested that a Southern Cattle Complex extending over many centuries organises the layout of settlements and kraals. In Europe, Treherne (1995) has suggested that a particular conception of the warrior band both changes over time and shows continuities from the Bronze Age into feudal political systems. Bradley (1990) suggests that the practice of hoard deposition, particularly in wet places, may have a long-term history in northern Europe. Other examples of the continuity of structures over the long term might include caste in India or geomancy in China. As in my own work on the long-term unfolding of the domus–agrios relationship in earlier European prehistory (Hodder 1990), it could be argued that all these examples of long-term continuities are constructed by us as observers in the present. And yet there are sufficient examples, sufficiently well argued, to suggest that some long-term routinisation of practices is demonstrable in the archaeological record. Indeed, as already implied above, if gradualism was not the major emphasis in cultural change, then typological seriation, one of the main building blocks of archaeological methodology, could not work.

As already argued, continuity of cultural practice and routine do not imply a lack of creativity. Fitting contingent events into existing schemes involves interpretive understanding and the search for creative solutions. Because of the predominance of interpretive understanding in long-term change, most of the creativity observed by the archaeologist is concerned with associational links and metaphor. Even in the creation of new nodes of information, there is a strong concern to make sense of novelty in terms of existing schemes. I would like to explore the idea that interpretive or associational creativity in humans dominates because we 'take in' more than 'give out'. Our minds seem geared primarily to absorb information rather than to create change actively – listen rather than speak, recognise and understand problems rather than solve them. While active and inventive creativity is perhaps the high point of human ingenuity, I would propose that much of the human mind is geared to interpretive creativity – that is, a fitting process rather than an active process of giving out information and generating radical change. Of course, the long process of learning in human maturation is presumably foundational in this emphasis on the absorbing of information.

1955

1957

1960

Figure 4.1 Gradual transformation of Cadillac car wings
Source: Time Earle and Christine Hastorf

1962

1965

1967

1968

The cognitive aspects of the selection and absorbing of information are presumably linked closely to the goals and intentions of the 'reader'. Understanding is not a passive cognitive process, but one aligned to the interests of the subject. The social basis for creativity is thus again underlined. And yet, the long slow processes observed in the archaeological record suggest that much creativity involves associational tinkering. This emphasis on mimicry, skeuomorphs, continuity and trend implies a predominance of forms of creativity linked to interpretive understanding. The social needs here vary from the avoidance of risk to the maintenance of power, but an underlying cognitive emphasis on understanding rather than invention seems plausible. From the archaeological point of view, I want to point out that the social nature of creativity seems to be reinforced by variation in degrees or types of creativity over time. A danger here is that creativity is simply equated with novelty and change. As already emphasised, the process of fitting and interpretive understanding involves creativity without necessarily leading to cultural change. Here the creativity may be involved in maintaining stability in the face of contingency.

A visit to the aceramic Neolithic site of Aşïklï Höyük in central Anatolia is a remarkable experience. In the eroded edge of the mound a street can be identified, always in the same place, endlessly replaced through time over centuries. In a deep sounding opened by Ufuk Esin (1991), perhaps 10 metres of deposit can be identified. In part of the deposit, the midden extends from the base to the top – centuries of depositing refuse in the same place, following the same routines. In another part of the sounding, buildings of exactly the same size are built on top of each other, walls exactly over earlier walls, the floors and hearths always in the same location. The routinisation of daily practices over centuries is remarkable when one considers that family sizes, wants and expectations must have changed through time. A similar, though less dramatic indication of continuity is seen at the later (seventh millennium BC) site of Çatalhöyük (Hodder 1996). Table 4.1 indicates the degree of continuity between levels in terms of the numbers of elaborate buildings in the same location from level to level. Both simple and elaborate buildings show remarkable continuity, especially in view of the fact that the excavations conducted by Mellaart (1967) did not cover the same area in each level. In general terms, the slow rate of change at these sites and in many Early Neolithic contexts replicates the long periods of cultural stability in the Palaeolithic. It is widely noted that the tempo of change increases through time into the present. This certainly suggests that creativity is heavily bound up with the social. It also suggests that forms of associational creativity may long have dominated over the creation of new nodes. Perhaps, taking a very long-term view, the possibilities for metaphorical association have increased exponentially because of greater information movement and exchange, greater complexity, etc. Perhaps the greater potential for associational links stimulates the potential

Table 4.1 The percentage of elaborate buildings in level n at Çatalhöyük that continue as elaborate buildings (a) from level n – l, and (b) to level n + l.

Level	*VIII*	*VII*	*VIB*	*VIA*	*V*	*IV*
(a)						
Number of elaborate buildings.		13	19	15	6	4
Number of elaborate buildings which continue.		2	7	14	3	2
Per cent		15	37	93	50	50
(b)						
Number of elaborate buildings.	4	13	19	15	6	
Number of elaborate buildings which continue.	2	7	15	5	2	
Per cent	50	53	79	33	33	

for nodal change. But it also has to be recognised that recent social and economic forms, such as capitalism, place a premium on the active generation of novelty – on creativity as part of the economic process. At Aşïklï Höyük and Çatalhöyük, at the other extreme, the social emphasis seems based on strong ties to lineage ancestors. In such a context, the emphasis is on creative linking rather than on creative difference.

CREATIVITY IN MATERIAL CULTURE

We could, however, also argue that the slow pace of change in material culture has something to do with the nature of material culture itself – that somehow material routines foster a particular emphasis on associational creativity and gradual change. This idea already occurred in the work of Pitt-Rivers (1874), when he distinguished an intellectual mind capable of reasoning on unfamiliar circumstances from an automaton mind capable of acting intuitively (Hodder 1989). On the one hand are found conscious reason and will, and on the other hand unconscious but still meaningful habits. For example, the child strives with the intellectual mind to learn how to walk or read or write, but this knowledge is then transferred to the automaton mind.

Pitt-Rivers was fascinated by the gradual transformation of form in ethnographic and archaeological artefacts. He looked at the way in which Iron Age forms gradually changed from their original design, and how axe forms changed gradually from the Palaeolithic to the Bronze Age. He argued that these gradual processes come about because artefact production becomes routinised within the automaton mind. The intellectual mind makes only minor adjustments leading to gradual stylistic 'progress'. The 'decay' of forms similarly comes about through, for example, the intellectual desire to save time or effort. Despite the fact that the overall sequence of forms is produced by the automaton mind, the intellectual mind is present, guiding the automaton mind to 'progress' or 'decay'.

This same distinction between conscious meanings and unconscious practices is found in archaeological authors throughout this century (Hodder 1989). For example, Crawford (1957) realised the importance of unconscious knowledge in prehistoric art or stone tool production. To make his point, he used an ethnographic analogy taken from Hutton's work among the Nagas of Assam. One day, Hutton observed a man chipping a pattern on the stone of a monument. He asked about the pattern's meaning, and the man replied that he did not know. The patterns were customary. Crawford concludes from this example that material forms can have meanings that are only 'dimly present' in the culture that produced them (ibid.: 68). Conscious knowledge of the meanings of artefacts may become lost, and yet the artefacts retain a meaning at the everyday, practical level.

The distinction between intellectual and automaton mind recalls that more recently described by Giddens (1979) between discursive and practical consciousness. The former refers to that which can be brought to and held in consciousness: knowledge that can be expressed. Practical consciousness refers to tacit stocks of knowledge that actors skilfully apply in conduct. It concerns knowledge about the use of rules, even if those rules cannot be formulated clearly in speech. In the work of Bourdieu (1977), practical or non-discursive knowledge is described as 'habitus': dispositions that are embodied. By this I mean that structural relations are inculcated within the body of the child as she grows up and moves around space. The bodily movements are how the child learns the relationships between people, their relative status and value. Moving around the house, people are 'put in their place'.

In the case of the work of Pitt-Rivers, it is as if creativity occurs only at the conscious level. Clearly it also occurs at the non-discursive level, as in the bodily movement that maintains one's balance on a bicycle. But because creativity at this level involves complex inter-relationships of mind/body routines, it may well be that fitting and linking is again the main concern, and thus long-term continuity and stability. These non-discursive links between mind and body may help to explain long-term continuities in material culture.

The complex inter-relationships of mind/body routines may extend across different domains of activity. I have often demonstrated this, at least to my own satisfaction, by conducting the following experiment in classes of students in Britain and the United States. Using the drawings in Figure 4.2, as yet un-named, I tell the students I am going to make two sounds. These are 'umanwama' and 'ikpikitik'. I then ask the students

Figure 4.2 Umanwama and Ikpikitik

to assign these two sounds to the two drawings. Always, no or only one or two (subversive!) students label the angular design 'umanwama'. I am not arguing for a cross-cultural pattern here, but it does seem remarkable that within a broadly defined culture, nonsense sounds that have never been heard before should be seen as appropriate for specific novel and nonsensical drawings. A cognitive link across domains is implied, between sound and sight. Many routines are domain specific, but this experiment implies that there may be a tendency to create resonances across domains. Indeed, the existence of such 'wholes' is assumed in much anthropological and archaeological method. There are, of course, many social and ecological advantages to such integration, from the pervasive influences of dominant ideologies to the ease of transfer of information. But there may be an underlying tendency for the human mind/body to create resonances across domains. Creativity in one domain thus has to contend with the search for resonances. This observation not only emphasises the argument that much creativity is about fitting. It also helps to explain the gradual and slow character of long-term change.

The link between mind/body routines and long-term change can perhaps also be explained in another way. The bodily routines invest labour in making artefacts, constructing monuments, establishing routes and pathways. People become tied to place within the situated practices of daily life (Barrett 1994; Thomas 1996; Tilley 1994). Such investing in the material world costs, and it establishes links between mind, body and place. Radical change becomes difficult, both practically, in terms of labour, and conceptually, in terms of the need to change routines. People become caught up in the material webs they create.

This is not just a material issue, however. It is also possible to argue that many human thought processes are dependent on things and words. As Geertz (1983) argued, culture is human nature. At least part of the reason for this is that the development of human cognition is based on the ability to symbol. The signified concept needs the signifier. The organisation of thought needs to be anchored in or pegged on to sounds, words and things. Of course we are most familiar with this dependence on objective symbols in relation to abstract thought. It is particularly in the construction of ideas and concepts that labels are needed, and particularly in the arena of communication. But it can also be argued that memory of a wide range of sensual and emotional experiences is tied to words and material cues.

Good examples of this latter characteristic of the mind/body are to be found in Proust's (1970 [1928]) novel *Swann's Way*, part of the sequence of novels entitled *Remembrance of Things Past*. The narrator in the novel describes taking some tea in which he had soaked a piece of a little cake called a 'petite madeleine' (p. 34). 'No sooner had the warm liquid, and the crumbs with it, touched my palate than a shudder ran through my whole body, and I stopped, intent upon the extraordinary changes that were taking place.' He describes a strong feeling of pleasure associated which he cannot at first make sense of in conscious thought. The madeleine has evoked some memory, in its form and taste, but this is initially a bodily memory that the conscious mind cannot understand or remember. The narrator tries and tries to call up the memory into consciousness. What was the madeleine recalling for him that gave such pleasure? Then

> suddenly the memory returns. The taste was that of the little crumb of madeleine which on Sunday mornings at Combray . . . , when I went to say good day to her in her bedroom, my aunt Leonie used to give me, dipping it first in her own cup of real or of lime-flower tea. (p. 36)

In that moment of bodily and then mental recognition,

> all the flowers in our garden and in M. Swann's park, and the water–lilies on the Vivonne and the good folk of the village and their little dwellings and the parish church and the whole of Combray and of its surroundings, taking their proper shapes and growing solid, sprang into being, town and gardens alike, from my cup of tea. (ibid.)

It is on processes such as that described by Proust that the use and re-use of monuments in prehistory (Bradley 1993) is presumably based. The re-use of monuments creates links in people's memories, and not just at the level of conscious communication. Places and monuments, objects and heirlooms, can evoke at the non-discursive level. The strategic use of place and history within social strategies manipulates remembrance at multiple levels. It is possible to do this because people are caught in the material webs of their experience and history.

But it is wrong to assume that the embedding of mind in a web of material symbols involves no creativity. As Proust's narrator tries to remember what the madeleine has evoked, the mind,

> the seeker, is at once the dark region through which it must go seeking, where all its equipment will avail it nothing. Seek? More than that: create. It is face to face with something which does not so far exist, to which it alone can give reality and substance, which it alone can bring into the light of day. (p. 35)

So memory and evocation, as much as they may work at the non-discursive level, are not mechanical processes – they involve selection and filtering in relation to the social and personal goals of the individual involved. Thus, while the non-discursive or automaton nature of much material culture practice may be linked to the gradualism of much material culture change, creativity is still involved. Creativity occurs in the selective memory of practices, as Proust suggests. But it also occurs through the process of trying things out and tinkering in practice. Ideas may flow from experimenting with things rather than from conscious theoretical musings. The ideas may flow indirectly, through lateral thinking or serendipity. Innovation often occurs when we just 'see what happens if I try this'.

In fact, in comparison to language, much material culture does change a lot and is sensitive to circumstance. It changes more quickly because it is more tied to practice and circumstance. While languages may continue largely unchanged over revolutionary social, political and economic change, some aspects of material culture at least will change, sensitive to context. But, as we saw with the example of innovation in aeroplane design, the practical and material and bodily concerns often lead to solutions being found that maintain things as they are. Creativity becomes a matter of linking in and making sense of things in established terms. The externality of words and materials limits creativity. The fact that our thought processes are embedded in the signifiers on which they come to depend, often at a bodily or non-discursive level, contributes to long-term continuities and also favours a form of creativity that focuses on fitting contingent events into the maintenance of routines.

RITUAL AS A MEDIUM FOR CREATIVE DEVELOPMENTS

If I am right about the predominance of associational, linking and fitting creativity in social life, then periods of major change must often be the cause of disruption of routine and must be dealt with in special ways. If the bodily routines and daily practices are to be realigned, for whatever reason, the new schemes may have to be introduced in peripheral or liminal areas before they can be adopted widely. Such a scenario implies that innovation is imposed or introduced, the problem simply being of achieving the widespread acceptance of change. However, it may often be in peripheral or marginal or liminal areas that creativity is enhanced. This may be because it is on the margins that routines can be questioned. It may be the case that marginal, subordinate groups most see the need for change and actively promote it. It may be that subordinate groups have less invested in traditional routines and have more to gain from change.

Ritual is often described as creating a liminal zone, outside space and time, outside the daily codes of conduct (Turner 1966). Indeed, in ritual, social relationships may be inverted or disrupted so that they can then be renewed. As such, ritual may often be involved in the maintenance of tradition, in the legitimation and reinforcing of the status quo. But I want to argue here that the very liminality and 'in-betweenness' of ritual also provide a moment for creativity and for the breaking of established schemes. It seems to be a common personal experience amongst those that describe creative moments that they 'let their minds go', 'think about something else', 'empty their minds', 'freely associate', 'let their minds go blank'. Whether it is a matter of creating new nodes or new links, the liminality and difference of ritual may provoke alternative ways of looking at the world. Thus ritual may not only promote the acceptance of change but also promote its creation.

In a number of instances in European prehistory, ritual sites or processes seem to predate non-ritual counterparts and major social change. For example, it can be argued (Hodder 1988) that Neolithic non-domestic enclosures, including those involving special activities, have a longer time span than defended or bounded occupation sites. The evidence is perhaps most clear in southern Scandinavia. At a number of sites such as Bjerggard, Toftum, Budelsdorf, Sarup, Trelleborg and Stavie, settlement traces continue after the initial uses of the enclosures for 'ritual' activity. The latter is indicated by special deposits in the ditches of complete pots, small heaps of tools, human jaws or, for example, dog skulls on a stone paving (Madsen 1988). However, at the tops of the ditches, above the ditches and in the interior there is often evidence of occupational use. In southern Scandinavia it is possible to argue that the idea of settlement agglomeration and communal centres first came about in a ritual context (Hodder 1988: 71).

A ritual context for major innovation is also seen in the formation of lineages in Neolithic northern Europe and in the introduction of the plough (ibid.). Cauvin (1978) argued that the domestication of cattle in the Near East derived from an initial symbolic domestication within ritual. Other, less well-documented examples include the following (Hodder 1988): the introduction of pottery in the Near East and the introduction of metallurgy in Europe, and the transition from theocracy to bureaucracy in many societies throughout the world.

CONCLUSION

I have argued in this chapter that discussions of creativity are problematic if they deal only with cognitive aspects. Instead, creativity needs to be understood in the context of mind/body. The creative process and the definition of creativity are undoubtedly historical and social. Creativity involves practical consciousness and the practices and routines of the body. It is worked at within the context of a mind/body search for resonances across different domains of life and experience. Play a Mozart piano sonata and then one by Beethoven. In the detail of the treatment and development of themes and in the details of the chord sequences one cannot help but wonder at the creative leap made by Beethoven. The discords, tension, anger and emotion so cleverly transform existing musical form. They seem minor in themselves but major in their accumulated effect. They usher in a new movement in music that we call 'Romanticism'. But the development of the Romantic spirit is part of a wider movement. The creativity demonstrated by Beethoven at least partly derives from resonances drawn with other spheres of life, whether new sensibilities about self and death (Tarlow 1992) or a new spirit of consumerism (Campbell 1987), or new social and political goals. Creativity, even of the most radical kind, is often largely about associational linking. The importance of this type of associational creativity is suggested by the archaeological evidence for a predominance of gradual and slow processes of change over the long term. I have argued here that the examples of major creativity that create new nodes may often result from backward looking re-evaluations of history in ideological terms. They may also result from a consideration of creativity at too short a distance. A longer-term view sees the specific and sudden in terms of the long and gradual. It sees an emphasis on creating links rather than creating differences.

Creativity occurs in the tension between routines and contingency. Such statements are of course very general, and in specific terms a wide range of diverse skills is involved in the general notion of creativity. Indeed, the latter term seems to be used whenever processes such as intelligent human adaptation lead to novelty and change. In describing creativity it has been necessary here to distinguish between different processes, such as interpretive understanding and goal-directed problem solving. But it remains difficult to distill out something distinctive as the 'creative process' from intelligence, imagination, adaptation, agency, problem recognition and problem solving. Music, art, engineering, archaeology all involve different creative skills. What is defined as creative changes historically. The concept dissolves. But, standing back, from an archaeological perspective, how it dissolves and how we creatively make it dissolve resonates with wider debates. We tinker and transform, within a long-term trend.

REFERENCES

Barrett, J. (1994) *Fragments from Antiquity*. Oxford: Blackwell.

Boden, M. (1990) *The Creative Mind*. London: Weidenfeld & Nicolson.

Bourdieu, P. (1977) *Outline of a Theory of Practice*. Cambridge: Cambridge University Press.

Bradley, R. (1990) *The Passage of Arms*. Cambridge: Cambridge University Press.

Bradley, R. (1993) *Altering the Earth*. Edinburgh: Society of Antiquaries of Scotland.

Campbell, C. (1987) *The Romantic Ethic and the Spirit of Modern Consumerism*. Oxford: Blackwell.

Cauvin, J. (1978) *Les Premiers Villages de Syrie–Palestine du IXème au VIIème Millenaire*. Lyon: Maison de l'Orient Méditerranéen Ancien.

Crawford, O.G.S. (1957) *The Eye Goddess*. London: Phoenix.

Esin, U. (1991) 'Salvage excavations at the pre-pottery site of Aşïklï Höyük in central Macedonia.' *Anatolica* 17: 123–174.

Flannery, K.V., and Marcus, J. (1983) *The Cloud People*. New York: Academic Press.

Geertz, C. (1983) *Local Knowledge: Further Essays in Interpretive Anthropology*. New York: Basic Books.

Giddens, A. (1979) *Central Problems in Social Theory*. London: Macmillan.

Hesse, M. (1995) 'Past realities.' In I. Hodder, M. Shanks *et al.* (eds) *Interpreting Archaeology*, pp. 45–47. London: Routledge.

Hobsbawm, E. and Ranger, T. (1983) *The Invention of Tradition*. Cambridge: Cambridge University Press.
Hodder, I. (1988) 'Material culture texts and social change: a theoretical discussion and some archaeological examples.' *Proceedings of the Prehistoric Society* 54: 67–75.
Hodder, I. (1989) 'This is not an article about material culture as text.' *Journal of Anthropological Archaeology* 8: 250–269.
Hodder, I. (1990) *The Domestication of Europe*. Oxford: Blackwell.
Hodder, I. (1993) 'Social cognition.' *Cambridge Archaeological Journal* 3: 25–37.
Hodder, I. (1996) *On the Surface*. Cambridge: McDonald Institute of Archaeology.
Huffman, T. (1984) 'Expressive space in the Zimbabwe culture.' *Man* 19: 593–612.
Latour, B. and Woolgar, S. (1979) *Laboratory Life: The Social Construction of Scientific Facts*. Beverly Hills: Sage.
Lechtmann, H. (1984) 'Andean value systems and the development of prehistoric metallurgy.' *Technology and Culture* 25: 1–36.
Lemonnier, P. (1989) 'Bark capes, arrowheads and Concorde: on social representations of technology.' In I. Hodder (ed.) *The Meanings of Things*, pp. 156–171. London: Unwin Hyman.
Lemonnier, P. (1993) *Technological Choices*. London: Routledge.
Madsen, T. (1988) 'Causewayed enclosures in south Scandinavia.' In C. Burgess, P. Topping, C. Mordant and M. Maddison (eds) *Enclosures and Defences in the Neolithic of Western Europe*, pp. 301–336. Oxford: British Archaeological Reports International Series, 403.
Mellaart, J. (1967) *Çatalhöyük*. London: Thames & Hudson.
Merriman, N. (1987) 'Value and motivation in pre-history: the evidence for "Celtic spirit".' In I. Hodder (ed.) *The Archaeology of Contextual Meanings*, pp. 111–116. Cambridge: Cambridge University Press.
Pitt Rivers, A. (1874) 'Principles of classification.' Reprinted in A. Pitt-Rivers (1906) *The Evolution of Culture*. London: Oxford University Press.
Proust, M. (1970) (first published 1928) *Swann's Way*. New York: Vintage Books.
Shanks, M. and Tilley, C. (1987) *Reconstructing Archaeology*. Cambridge: Cambridge University Press.
Tarlow, S. (1992) 'Each slow dusk a drawing down of blinds.' *Archaeological Review from Cambridge* 11: 125–140.
Thomas, J. (1996) *Time, Culture and Identity: An Interpretative Archaeology*. London: Routledge.
Tilley, C. (1994) *The Phenomenology of Landscape*. London: Berg.
Treherne, P. (1995) 'The warrior's beauty.' *Journal of European Archaeology* 3: 105–144.
Turner, B. (1984) *The Body and Society: Explorations in Social Theory*. Oxford: Blackwell.
Turner, V. (1966) *The Ritual Process*. London: Routledge & Kegan Paul.
van der Leeuw, S. (ed.) (1989) *What's New*. London: Unwin Hyman.
von Gernet, A. and Timmins, P. (1987) 'Pipes and parakeets'. In I. Hodder (ed.) *Archaeology as Long Term History*. Cambridge: Cambridge University Press.

CHAPTER FIVE

Creative thought in traditional Aboriginal society

ROBERT LAYTON

The anthropological study of creative thought in traditional Aboriginal society is problematic. If 'traditional Aboriginal society' means indigenous society in Australia prior to the colonial period, then the subject is unknowable to anthropology. It is always possible to reject the validity of evidence on creativity from contemporary Aboriginal society on the grounds that society has been radically transformed by the unprecedented impact of colonialism. Swain has recently used a similar argument to reject evidence that runs counter to his reconstruction of Aboriginal religious philosophy prior to colonisation (Swain 1993). There are two ways of approaching the problem. One is to study creativity in contemporary communities, in the way that demonstrably indigenous themes are realised, and to argue that creative responses to colonisation have developed by extension from traditional practice. The other approach is to look for archaeological evidence of change in the past expressions of culture, and to draw inferences about the processes that made such creativity possible.

CREATIVITY IN CONTEMPORARY INDIGENOUS COMMUNITIES

The conceptual space of traditional culture in contemporary Australia is bounded by a creation period, often referred to in popular writings as 'The Dreamtime'. During the creation period, indeterminate possibilities became determined through the actions of heroic beings who left their mark upon the landscape and gave society its structure. Choices were made between death and regeneration, social obligations were established, upheld or denied, with perpetual consequences. The events of the creation period are retold in legend. In analysing how novel structures were created within this apparently fixed, encompassing space, I shall draw on Margaret Boden's ideas (Boden 1994), as well as on my own work on creativity in art (Layton 1981: ch. 5, 1992: 115ff.).

A common theme in northern Australia is that the ancestors left the spirits of unborn children in pools of permanent water along their track. Such places along creeks and billabongs are now sacred sites, protected by the clans who 'own', or are responsible for looking after, each stretch of country (or clan 'estate'). At death, the clansperson's spirit returns to the water from which it came. The travels of the totemic ancestors are re-enacted in rituals at which living members of the clan take on the *personae* of the ancestors. The participant's distinctive body paint conveys the dual human–animal essence of the ancestor and, like the heraldic devices of medieval Europe, asserts the actor's right to membership of the clan and its estate. In central Australia, traditional ceremonies are performed around ground paintings that depict the ancestors' travels. Frequently, the ancestor's footprints are shown passing from site to site.

Each telling of a legend, each fresh realisation of a painted composition and each performance of ceremony necessitates creative decisions by the narrator or the managers of the performance. The ambiguity inherent in attitudes to the dead can be resolved in more than one way, depending on how immediate experiences are likened to prototypical events. In the 1920s, for example, Lloyd Warner described how the Yolngu of North-east Arnhem Land gave conflicting accounts of the nature of a dead person's spirit. While believing, in principle, that each person left two spirits, the ancestral *birimbir* and the trickster *mokoi*, some said that the spirit's identity was in doubt until his or her clan song cycle had been performed (Warner 1937: 413–415). The *mokoi* is sometimes said to hang around the burial platform, so people usually pass at a distance. As they pass, 'the men, when they are with the women, sometimes tease them by grunting and coughing like *mokois*, and laugh when the women jump. The women laugh too, but usually not until later' (ibid.: 433). In the 1980s, Morphy showed how the series of songs performed by the Yolngu to transport the deceased's *birimbir* back to the clan well are put together anew for each funeral, depending on the place of death, the possible routes the spirit might take across other clans' countries to reach its clan waterhole, and which of these clans are represented at the funeral (Morphy 1984: 87ff). Venbrux has recently published a detailed analysis of a funeral on Melville Island, off the coast of Western Arnhem Land. He writes,

> In the rituals the performers or narrators fit their own stories within the overall framing story. . . . The stories 'told' by the performers . . . [are] related to current happenings in their social life. As these stories run through the lives of the narrators, they help them shape their culture and adjust to new situations
>
> (Venbrux 1995: 141)

Elsewhere, I have quoted examples of different tellings of the same legend from the Kimberley region of Northwest Australia, and argued that such creative retelling is inherent in the structure of Aboriginal cognition (Layton 1992: 40–45, 116). Although Margaret Boden describes the novel combination of familiar ideas as the simplest and least original form of creativity (Boden 1994: 520–521), there is no doubt that the grammars of Australian myth and ritual routinely enable the creation of new, but always possible, 'sentences'.

Aboriginal politics also have an essential strand of creativity, resolving indeterminacy in how the ancestral law should be upheld. If a clan is threatened with extinction through the failure to bear children, others must be trained to succeed to their responsibilities. Decisions must be taken about the most appropriate successor, and who should train them. Social identity is acknowledged to be an arena for indigenous political contestation, and any individual's claims to knowledge of the ancestral order are subject to political assessment. The ancestral framework within which these negotiations are conducted is, however, considered to be unchanging. Marriages that fail to follow the prescribed pattern also necessitate negotiation of the children's status. Among the Alawa of the Gulf Country, those who inherit membership of the group responsible for an estate or country through their fathers are termed *miniringgi*. Those who inherit membership through their mothers are termed *junggaiyi*. Those who stand in the relationship of *miniringgi* to a ceremony celebrating creation of the country during the ancestors' travels must ask for it to be performed, but it is the *junggaiyi* who decide on timing. *Junggaiyi* prepare the ceremonial dance ground and equipment, they decorate the *miniringgi* and sing the songs tracing the ancestors' routes from place to place. The interdependence of *miniringgi* and *junggaiyi* facilitates consistency in performance as knowledge is passed on from one generation to the next.

Intra-moiety marriages are strongly discouraged. If a man's father and mother come from the same patri-moiety, he is potentially both *miniringgi* and *junggaiyi* to the same ceremonial complexes. The two roles are absolutely opposed. *Junggaiyi* must perform duties forbidden to *miniringgi*. A decision has to be made as to which role will be chosen before the individual can take on ceremonial status. Alawa discourse provides alternative propositions. Conventionally, children of wrong marriages are assigned the semi-moiety and subsection status they would have received had their mother married correctly. If the father's family are powerful, however, they can insist that the children 'follow the father'. Whichever course of action is taken, one group will lose potential members. The outcome of any case will be a matter for negotiation (see Bern 1979a), and powerful arguments can be mounted on either side. Where a person's spirit originates in a water outside their father's estate, it is also possible to negotiate their membership of the group owning the estate of conception. The cultural system thus provides a syntax of social relationships. Ambiguity is eliminated by assigning persons to positions specified by the system mapped out upon the landscape by the ancestors' travels (for an example from central Australia, see Layton 1995: 223–230).

While this process might not be considered truly creative, it demonstrates the way in which Aboriginal people can move around in, and explore, their conceptual space, a skill that Boden argues is a prerequisite to transforming that space (Boden 1994: 523). Linguistic evidence suggests that the kinship system used by the Alawa and many neighbouring communities was developed by combining two simpler systems (McConvell 1985). This transformation doubled the number of kinship positions in the system and made possible a third ritual role, that of *darlnyin*, which one plays toward the clan of one's mother's mother and her brother. Under the simpler systems, these roles would have been indistinguishable from those of father's father and his sister. *Darlnyin* are particularly charged with ensuring that *miniringgi* wear their own clan's body painting, and do not inadvertently appear to claim another clan's land by bearing their painting.

THE POLITICAL ENVIRONMENT OF CREATIVITY

The work of writers such as Morphy (1984, 1991) and Venbrux (1995) provide rich examples of creativity in the routine realisation of Aboriginal culture. If the pace of cumulative change in Australia's indigenous cultures appears slow when compared, for example, to European art of the past few millennia, the reason must be found in the social environment into which creativity is manifested rather than in the cultural 'engine' generating creativity. Munn demonstrated that women's art among the Warlpiri of central Australia was at least as creative as men's art, since women were often inspired during dreams to create new variants on existing compositions. Paradoxically, however, the existing corpus of women's motifs was smaller than that in men's art. Munn argued that the men's clan-oriented cult art provided a forum in which new variants could be adopted, regularly performed and carefully memorised within a corporate group, whereas women's cults were performed less frequently and by more variable, contingent sets of participants (Munn 1973: 40–41). Bern also noted that men's religious cults constitute the most elaborated aspect of Aboriginal culture. Yet, he argued, the segmentary structure of men's cults, in which members of each clan held an exclusive body of knowledge and designs, orally transmitted, itself imposed a limit on the extent to which social differentiation can accumulate. Each clan's knowledge constitutes a segment of the totality associated with the tracks of ancestors who passed through its territory. Men's authority is built upon reciprocal dependence, both in the exchange of marriage partners and in the performance of ceremonial roles such as *miniringgi*, *djunggaiyi* and *darlnyin* (Bern 1979b: 126–127). Ritual interdependence also helps to meet the practical need for mutual access to the subsistence resources of other clans' estates, to cushion local variability (Layton 1986a: 257, 1995: 227–228). It is within this stable political environment of mutual dependence that creative innovations must become institutionalised if they are to persist. No clan could afford to diverge too far from the patterns sustained by the wider society.

RESPONSES TO COLONISATION

The impact of colonisation subjected indigenous communities to wholly new experiences. The way in which sense was made of these experiences reveals indigenous creativity taken to its limits. Radical change in the political environment also enabled the institutionalisation of new traditions.

Many leading anthropologists have argued that the cults that are performed over much of the 'Top End' of the Northern Territory this century were devised in response to colonisation. There is no doubt that they were spreading during the colonial era. The spread of the G.[1] is well documented (Berndt 1951: xvii–xx; Elkin 1952: 251, 1961: 167–169; Meggitt 1966; Stanner 1963: 244–245) as is the spread of the Y. (Bern 1979a: 48). All of the key elements of the Gulf Country G. were seen to be duplicated by Ronald Berndt in 1949, in a performance in Northeast Arnhem Land (Berndt 1951: 40–47). Berndt argued that the cult had spread into Northeast Arnhem Land from the Roper River (ibid.: xxv). He noted that the heroes celebrated in the Gulf Country G. are associated with a hole in a creek on Alawa or Mara Country near Hodgson Downs Station (ibid.: 148). Elkin considers that the Y. also developed in the middle and lower Roper Valley (Elkin 1961: 167), close to Hodgson Downs and the centre of early colonial conflict.

It is therefore possible that new or modified religious cults were introduced to assert pan-Aboriginal traditional claims to land ownership in the face of attempted dispossession by pastoralists, providing an example of Boden's "transformations of conceptual space" (Boden 1994: 523). One of the most interesting aspects of these cults is the way in which the travelling heroes whom they celebrate interact with local heroes who assist or challenge them. What renders each clan's body of tradition unique is not the identity of the travelling ancestors, but the identity of the local ones belonging to its estate. It is tempting to speculate that the travelling figures constructed a sense of regional unity out of previously more self-contained clan traditions. Even if this is so, however, the regional cults may have originated prior to colonial intervention. Both Y. and G. cults were recorded by the earliest anthropologists to work in the Gulf Country, Spencer and Gillen, who learned of them in 1902 (Spencer and Gillen 1904: 223), only 20 years after the introduction of cattle ranching. Spencer and Gillen's account of the Alawa's southern neighbours, the Nganji, shows that the cultural system is essentially unchanged since the turn of the century. It is possible that the cults already existed locally or that they developed from existing, pre-colonial cults (I will refer later to one line of evidence for the antiquity of some). Today, they are nonetheless performed as an explicit assertion of rights to land against the colonists. As one senior Alawa man put it to a white Australian lawyer I was working with in 1994: 'If we didn't do that ceremony, someone like you mob might shoot us and drive us off our land.'

REPRESENTING COLONISATION IN A TRADITIONAL IDIOM

Associated with the cave at Walgundu, close to Alawa Country, is a legend that a band of people once encountered a large snake. Unaware that the snake was a heroic being, they killed it. In revenge, the snake's friend Lightning drove the people into Walgundu and killed them there (see Hill 1951: 361; Holmes 1965: 11; Miles 1986: 269). When she visited Walgundu in the early 1960s, Sandra Holmes saw a number of skeletons, some of which appeared to have bullet wounds, suggesting another interpretation of the cause of death (Holmes 1965: 11). Suppose Walgundu is indeed a massacre site; how would the event be related in an indigenous discourse of the initial period of colonisation?

During research for the Cox River land claim, Willy Gudabi told me how his Alawa forebears had reacted after first seeing cart tracks on the old stock route of the 1870s, from Queensland to the Northern Territory. In those days, people had never seen a wheeled vehicle, and they thought the tracks left where the stock route crossed the Hodgson River were those of an enormous snake. In fear, they fled to the hills to hide. In 1896, Archibald Meston reported on the harassment both pastoralists and fishermen had inflicted on the Aboriginal people of North Queensland. One old man to whom Meston spoke remembered the Kennedy expedition of 1818, whose members had shot numerous Aboriginal people they sighted. Meston was told that the terrified inhabitants 'thought guns were the source of thunder and lightning' (Meston 1896: 3). These reminiscences make it possible to see how the totally unprecedented experience of a massacre might be represented, in a traditional idiom, as an alliance between ancestral beings who punish a breach of traditional law. Today, however, Alawa people can readily conceptualise and relate both the impact of massacres and Aboriginal resistance.

A rather similar case in the Western Desert of central Australia arose through an encounter between a gold prospector, Henry Lasseter, and the local Pitjantjatjara people. Lasseter's camels had escaped with his supplies and he hid in a rock shelter, begging the local people to take a message to the nearest cattle station, some 250 kilometres to the east, but refusing to eat the bush food they brought him. At one point Lasseter fired his revolver at a group of curious young people accompanying an old man. According to Lasseter's diary they 'ran like rabbits'. Paddy Uluru, who would have been a young man (although not an eyewitness) when this event took place, told me the old man was a ngangkari (native healer or doctor) and the bullets bounced harmlessly off a stone in his chest (see Layton 1986b: 57).

Debbie Rose has recorded a legend current in the Victoria River district that Ned Kelly, the famous Irish-Australian outlaw, came to Australia before Captain Cook and befriended Aboriginal people. Captain Cook, on the other hand, stole Australia and its minerals from the indigenous inhabitants (Rose 1992: 195–199). Although, as Rose shows, the legend is expressed in the discourse of the Creation Period it has, nonetheless, creatively absorbed and transformed motifs from colonial legend. Venbrux documents a similar process within a traditional context, in which a Tiwi murder that occurred around the turn of the century has taken on legendary resonances (Venbrux 1995: 45–56).

TRADITIONAL ART IN CONTEMPORARY AUSTRALIA

The opportunity to sell paintings on a market created by the colonists provoked a substantial restructuring of what was regarded as possible and acceptable in artistic production. I will briefly illustrate this with two examples from northern and central Australia.

In Western Arnhem Land, the market for Aboriginal bark paintings was created by Baldwin Spencer, who visited the area in 1913 and bought a large collection of bark paintings through the agency of a local buffalo hunter. During the 1950s, Ronald and Catherine Berndt promoted Aboriginal culture by arranging the exhibition of paintings from the area (Jones 1988: 174). It is now a well-known centre for commercial art. Commercial paintings have clear links with the rock art of recent centuries, but frequently use novel compositions that combine the recent X-ray art with the older 'dynamic' tradition of human figures engaged in hunting and gathering. The depiction of plants, which is rare in pre-colonial rock art, has also become more common.

Morphy describes how Yolngu clans in Northeast Arnhem Land made different decisions about which legendary themes to release for sale on commercial paintings. He also shows how the 'grammar' of traditional design allows a wide variety of different compositions to be generated around a single theme (Morphy 1991: ch. 9). Munn has shown that central Australian art also has a 'grammar' of 'core' and 'adjunct' motifs that allows a range of compositions, the potential of which has been explored in new ways on the large canvases painted since the development of commercial art. Pintupi, Arrente and Lurija artists at Papunya, in central Australia, were faced with similar decisions to the Yolngu which they solved by using traditional motifs and compositional principles, but omitting details that could only be produced in secret contexts. In Margaret Boden's terms, these are generative systems (cf. Layton 1981: ch. 5).

THE EVIDENCE FROM ARCHAEOLOGY

Had any change as radical as that wrought by colonisation ever happened before? Few events could have had comparable impact, but it is possible that the rising post-glacial sea levels demanded a similar degree of creative response.

Rock art has been created for over 10,000 years; possibly up to 30,000 years (see Flood 1996; McDonald *et al.* 1990; Nobbs and Dorn 1993; Taçon and Brockwell 1995). A comparison of central and northern Australia shows that central Australian art appears to be less subject to creative innovation than that of the north. I do not wish to argue that central Australian cultures are inherently more conservative: attachment to the land is more flexible in the centre, and provides more scope for the creative construction of social allegiances. Rather I wish to argue that relative opportunities for creative variation are inherent in the two styles, that one is a relatively open system, the other more closed. In northern Australia, there are many rock art motifs depicting elements of colonial culture. Stencils of traditional artefacts such as boomerangs are joined by those of introduced artefacts such as steel axes. Horses and donkeys are represented using forms clearly derived from the traditional kangaroo motif: the hind legs are noticeably longer than the forelimbs, but the feet are shown as horse-shoes rather than kangaroo toes. In Western Arnhem Land, X-ray ships are added to panels already painted with X-ray fish. The process of extension and modification corresponds to Boden's account of computer programs that model artistic creativity, in which successful rules are 'selected for "breeding" future generations' of artwork (Boden 1994: 528, and this volume)

In central Australia, recent rock art and commercial paintings use essentially the same vocabulary of motifs as the ancient Panaramittee tradition, which Nobbs and Dorn argue is up to 30,000 years old (Nobbs and Dorn 1988). There is, however, evidence that the artistic system has changed over time. Munn identified the typical compositional principles that generate recent, traditional ground drawings as those of the 'site+path' and 'core+adjunct'. The 'site+path' composition depicts the track of an ancestral being moving between sets of concentric circles that represent places. The 'core+adjunct' juxtaposes a relatively generalised core motif, such as a wavy line, with peripheral forms that enable it to be identified as the trace left by, for example, a snake, possum's tail or dragging spear (Munn 1973). These compositional principles can also be identified in recent rock art, and have been extended to commercial art, but are not seen in the ancient Panaramittee rock engravings. Elsewhere I have argued that other changes in central Australian rock art suggest that it became incorporated into a totemic system several thousand years ago (Layton 1992: 236). The increasing complexity in the art's visual grammar may be a related process.

Central Australian art makes 'reference' to the world of objects primarily by depicting the traces people, animals and artefacts leave in the sand. The simplicity of the motifs enables each to be read in various ways and each can thus have many levels of meaning. Munn showed how concentric circles can represent a camp, a sacred waterhole (itself the transformed remains of an ancestor's camp) and the creative power of the ancestor emanating from the site (Munn 1964). An arc typically represents the mark impressed on the ground by a seated person. A pair of concentric arcs represent the boomerangs used as clapsticks to beat the rhythm during a ceremony. By extension, pairs of arcs may therefore represent ancestors sitting in their camp performing the prototypical ceremony that now commemorates them. The 'creative urge' of the tradition is directed toward perceiving ever more layers of congruent meaning behind a limited stock of essentially ambiguous figures, rather than developing an increasing vocabulary of motifs. Whereas introduced animal species are widely depicted in the art of northern Australia, no definite examples of the footprints of introduced species have been reported in geometric art.

In Western Arnhem Land, however, even the most ancient art reveals a concern for detailed iconography. The depiction of extinct animals such as the thylacine is one of the best lines of evidence for its antiquity. This stylistic tradition makes reference to the world in a different fashion, which better suits it to recording change. The oldest tradition is one of lively compositions. A later phase, however, sees the appearance of two new motifs, the 'Yam Figure' and the Rainbow Serpent (see Taçon and Brockwell 1995; Taçon *et al.* 1996). The meaning of the Yam Figure is lost to local culture, although a Yam ritual is performed on the Tiwi islands off the coast of Western Arnhem Land during the monsoon season (Venbrux 1995: 119ff.; cf. Lewis 1988: 109). On the mainland, the Rainbow Serpent is today associated in legend with a catastrophic flood that led to the creation of a society based on totemic clans. The analogy (cf. Boden 1994: 525) between snake and rainbow is a central image in these legends. One of the most interesting lines of evidence for the antiquity of modern religious cults is that initiates in the 'Rainbow Serpent' cult of the Yolngu, a variant of the G. cult described by Warner, are said to be like flying foxes, which also appear in early Rainbow Snake paintings (see, for example, Chaloupka 1993: plate 145). Daryl Lewis has suggested that the appearance of the Rainbow Serpent is evidence for new cults precipitated by the crisis of rising sea levels (Lewis 1988: 91). Both Lewis and I have put forward similar arguments concerning the Wanjina paintings of the Kimberleys, which replace an earlier, dynamic style (Layton 1992: 236; Lewis 1988: 109). Taçon and Chippindale have recently argued that the appearance of scenes of combat in Western Arnhem Land also expresses the rising tensions created by pressure on land (Taçon and Chippindale 1994). Regional cults resembling the modern G. and Y. may have originated at this time.

The recent 'X-ray' art of Western Arnhem Land is no more than 2,000 to 3,000 years old. Lewis, Taçon and others have pointed out that, whereas the oldest, 'Dynamic' rock art of the region depicts *land* animals, the majority of X-ray motifs are of the fish and other species hunted today on the black soil plains that developed after the post-glacial sea rise stabilised (Taçon 1988).

Thus, in conclusion, it can be argued that there is both a strong, conservative strand in traditional Aboriginal society, which has enabled the impact of colonialism to be withstood, and for traditional rights to land to be asserted in a traditional idiom, and also a creative strand which repeatedly generates new variants of cultural practices and, more rarely, transforms the cultural structure itself. The segmentary, relatively egalitarian character of traditional Australian societies normally determines the fate of creative variants. Occasional catastrophic change, inflicted by processes such as rising sea levels or colonial conquest has sometimes provoked radical change, in which creativity has been tapped as a means of reflecting upon, rebuffing and taking advantage of, the new natural or social environment.

NOTE

1 In deference to Aboriginal insistence that the names of these cults should not be spoken in public, I refer to them by their first letters.

REFERENCES

Bern, J. (1979a) 'Politics in the conduct of a secret male ceremony.' *Journal of Anthropological Research* 35: 47–60.

Bern, J. (1979b) 'Ideology and domination: toward a reconstruction of Australian Aboriginal social formation.' *Oceania* 50: 118–132.

Berndt, R.M. (1951) *Kunapipi: A Study of an Australian Aboriginal Religious Cult.* Melbourne: Cheshire.

Boden, M. (1994) 'Précis of *The Creative Mind: Myths and Mechanisms.*' *Behavioural and Brain Sciences* 17: 519–531.

Chaloupka, G. (1993) *Journey in Time: The World's Longest Continuing Art Tradition.* Chatswood, NSW: Reed.

Elkin, A.P. (1952) 'Cave paintings in southern Arnhem Land.' *Oceania* 22: 245–255.

Elkin, A.P. (1961) 'The Y. . . .' *Oceania* 31: 166–209.

Flood, J. (1996) 'Culture in early Aboriginal Australia.' *Cambridge Archaeological Journal* 6: 3–36.

Hill, E. (1951) *The Territory.* Sydney: Angus & Robertson.

Holmes, S. le B. (1965) 'The Nagarren Y. . . .' *Milla Wamilla, The Australian Bulletin of Comparative Religion* 5: 8–17.

Jones, P. (1988) 'Perceptions of Aboriginal art, a history.' In P. Sutton (ed.) *Dreamings: The Art of Aboriginal Australia*, pp. 143–179. New York: Braziller.

Layton, R. (1981) *The Anthropology of Art* (first edition). St Albans: Granada.

Layton, R. (1986a) 'Political and territorial structures among hunter gatherers.' *Man* (n.s.) 21: 18–33.

Layton, R. (1986b) *Uluru, an Aboriginal History of Ayers Rock*. Canberra: Aboriginal Studies Press.

Layton, R. (1992) *Australian Rock Art: A New Synthesis*. Melbourne: Cambridge University Press.

Layton, R. (1995) 'Relating to the country in the Western Desert.' In E. Hirsch and M. O'Hanlon (eds) *The Anthropology of Landscape*, pp. 210–231. Oxford: Clarendon.

Lewis, D. (1988) *The Rock Paintings of Arnhem Land: Social, Ecological and Material Culture Change in the Post-glacial Period*. Oxford: British Archaeological Reports S415.

McConvell, P. (1985) 'The origin of subsections in northern Australia.' *Oceania* 56: 1–25.

McDonald, J., Officer, K., Jull, T., Donahue, D., Head, J. and Ford, B. (1990) 'Investigating ^{14}C AMS: dating prehistoric rock art in the Sydney Sandstone Basin, Australia.' *Rock Art Research* 7: 83–92.

Meston, A. (1896) 'Report on the Aboriginals of Queensland.' *Queensland Legislative Assembly Votes and Proceedings* 4: 17.

Meggitt, M.J. (1966) *Gadjari among the Walbiri Aborigines of Central Australia*. Sydney: Oceania publications, monograph 14.

Miles, B. (1986) 'Rose red city.' *Journal of the Sydney Speliological Society* 30: 267–273.

Morphy, H. (1984) *Journey to the Crocodile's Nest*. Canberra: Aboriginal Studies Press.

Morphy, H. (1991) *Ancestral Connections: Art and an Aboriginal system of Knowledge*. Chicago: University of Chicago Press.

Munn, N. (1964) 'Totemic designs and group continuity.' In M. Reay (ed.) *Aborigines Now*. Sydney: Angus & Robertson.

Munn, N. (1973) *Walbiri Iconography: Graphic Representation and Cultural Symbolism in a Central Australian Community*. Ithaca: Cornell University Press.

Nobbs, M. and Dorn, R. (1988) 'Age determination for rock varnish formation within petroglyphs: cation ratio dating from the Olary Region, South Australia.' *Rock Art Research* 5: 108–146.

Nobbs, M. and Dorn, R. (1993) 'New surface exposure ages for petroglyphs from the Olary Province, South Australia.' *Archaeology in Oceania* 28: 18–39.

Rose, D. (1992) *Dingo Makes us Human: Life and Land in an Australian Aboriginal Culture*. Cambridge: Cambridge University Press.

Spencer, W.B. and Gillen, F.J. (1904) *The Northern Tribes of Central Australia*. London: Macmillan.

Stanner, W.E.H. (1963) 'Cosmos and society made correlative.' *Oceania* 33: 239–273.

Swain, T. (1993) *A Place for Strangers: Towards a History of Australian Aboriginal Being*. Cambridge: Cambridge University Press.

Taçon, P. (1988) 'Identifying fish species in the recent rock paintings of Western Arnhem Land.' *Rock Art Research* 5(1): 3–15.

Taçon, P. and Brockwell, S. (1995) 'Arnhem Land prehistory in landscape, stone and paint.' *Antiquity* 69: 676–695.

Taçon, P. and Chippindale, C. (1994) 'Australia's ancient warriors: changing depictions of fighting in the rock art of Arnhem Land.' *Cambridge Archaeological Journal* 4: 211–248.

Taçon, P., Wilson, M. and Chippindale, C. (1996) 'Birth of the Rainbow Serpent in Arnhem Land rock art and oral history.' *Archaeology in Oceania* 31: 103–124.

Venbrux, E. (1995) *A Death in the Tiwi Islands*. Cambridge: Cambridge University Press.

Warner, W.L. (1937) *A Black Civilisation*. Chicago: University of Chicago Press.

PART II

THE EVOLUTION OF HUMAN CREATIVITY

CHAPTER SIX

INTRODUCTION TO PART II

H. sapiens sapiens, the species you and I belong to, is unique. All species are unique – as Rob Foley (1987) explained in his provocatively titled book about human evolution, we are just *Another Unique Species*. But we are unique in a spectacular fashion. We belong to the only genus with a single member. We live in a wider diversity of environments than any other species, ranging from deserts to cities to space stations. We have language, are prone to believe in supernatural beings, create weapons of mass destruction and question the meaning of our existence. By these characteristics we are unique in a far more profound way than any other species is unique.

Of all these features it is the first – being the only living member of our genus – that is perhaps the most significant but the least recognised. Discoveries of fossils made during the last two decades have shown that throughout human evolution up until a mere 30,000 years ago, there was more than one type of human alive at any one time. During the early stages we can recognise at least three species that appear to have substantially overlapped in time: *H. habilis*, *H. rudolfensis* and *H. ergaster*. At later stages, we are able to identify at least five species – *H. erectus*, *H. heidelbergensis*, *H. neanderthalensis*, archaic *H. sapiens* and *H. sapiens sapiens*, some of whom may have significantly overlapped with each other in time and space. In effect, it is now clear that the pattern of human evolution is like a bush with many branches, rather than a ladder with a step-wise 'progression' from species to species (Figure 6.1). It is quite anomolous that for the past 30,000 years there has only been one member of our genus alive.

Why is that the case? Competitive exclusion appears to be the answer. The most likely scenario for modern human origins is that there was a single origin in Africa between 200–100,000 years ago, from where they dispersed across the globe, gradually replacing all other types of humans (Stringer and McKie 1996). This replacement is unlikely to have been by violence, but simply by being more efficient at exploiting the environment, more able to cope with environmental extremes, and more flexible in social behaviour. These achievements can be attributed, in part, to

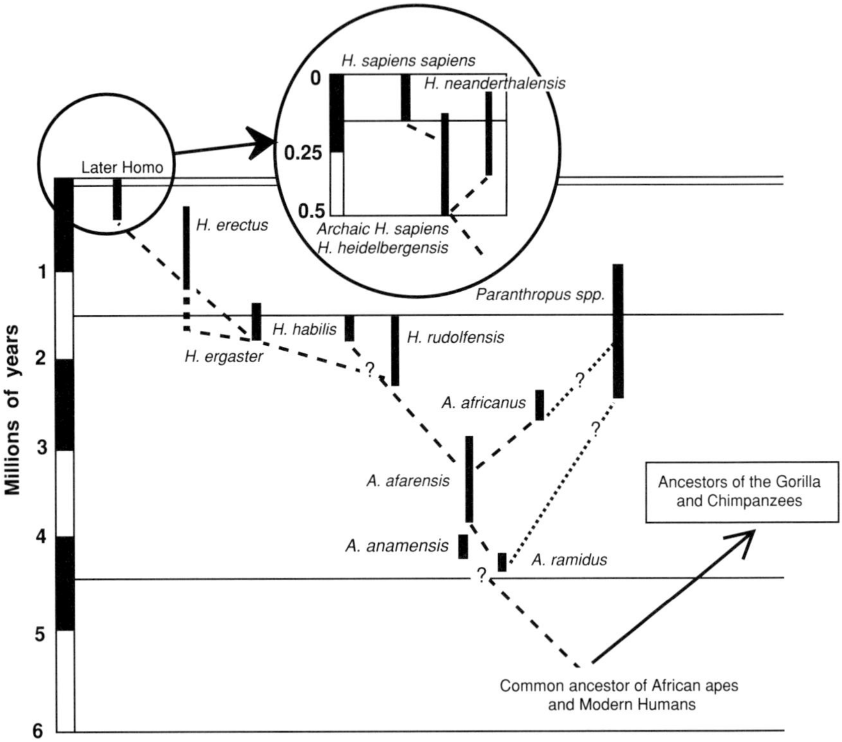

Figure 6.1 Human evolutionary tree. This is just one possible version of the evolutionary tree that can currently be produced with known fossils and dates. Other trees may include more or fewer species, and present these in different evolutionary relationships

the use of material culture in a fundamentally different way to that of existing hominid species – a use that involved artefacts to actively structure interactions with the social and natural worlds. Whether this was, in turn, simply due to discoveries and inventions made by modern humans – which might equally have been made by other types of humans – or due to a different type of mentality, so that only modern humans could have possibly used material culture in this fashion, is an issue hotly debated with archaeology and palaeoanthropology.

If we are dealing with a different type of mentality, as I firmly believe (Mithen 1993, 1996), then the possession of modern language is an obvious candidate for 'what made the difference' (Mellars 1991; Whallon 1989). But so too is the possession of a particular creative intelligence – which may, or may not, be dependent itself upon language, as will be discussed later in this book.

We have seen in the first contributions to this volume that creativity has several elements, one of which is the ability to explore conceptual spaces partly through the use of analogy and metaphor. Creative thinking can be devoted to both solving and recognising problems, to generating as well as interpreting action. In this part of the book, we ask about the evolution of human creativity – we ask whether we can see traces of this in our early ancestors and relatives, or whether we should attribute creative thinking to modern humans alone. We ask how and when creative thinking arose during evolution.

In this regard, this section of the book is following the proposition made forcefully during the last decade that any feature of the modern mind can be fully understood only when placed into an evolutionary context. This has happened most clearly with language, with a remarkable flurry of books, papers and conferences about its evolution (e.g. Aitchison 1996; Bickerton 1996; Corballis 1992; Dunbar 1996). Other edited collections and single-authored books have explored the evolution of the human mind in more general terms (e.g. Barkow *et al.* 1992; Corballis, in press; Cummins and Allen, in press; Mithen 1996; Noble and Davidson 1996). There has been limited attention within all of these works to human creativity.

This section has four chapters: one devoted to the thought of the common ancestor of humans and apes, *c.*6 mya (million years ago), one to the earliest members of the *Homo* genus, *c.*2 mya, one to Neanderthals. *c.*200–30,000 years ago, and one to modern humans at the time of the Middle/Upper Palaeolithic transition, *c.*30,000 years ago. As such, this section has enormous gaps in the story of human evolution and with regard to the evidence that may pertain to the evolution of creative thought. But these four chapters address key turning points and key species in our quest to understand human evolution. To appreciate these chapters, we must first situate their topics within the larger picture of our evolutionary past.

HUMAN EVOLUTION

The last decade has seen remarkable developments in our understanding of human evolution – not just in the 'facts' of the story, but also in the ways in which those 'facts' are established, and indeed what questions we want to ask about our evolutionary past. As evidence, one can simply consult the pages of the *Cambridge Encyclopedia of Human Evolution* (Jones *et al.* 1992) and appreciate the diversity of disciplines and approaches it contains.

In terms of the discovery of fossilised remains, there have recently been particularly important finds from east Africa with regard to the earliest hominids – *Australopithecus ramidus* (*c.*4.5 mya) and *Australopithecus anamensis* (*c.*4.2–3.9 mya) (White *et al.* 1994; Leakey *et al.* 1995). The discovery in 1984 of an almost complete *H. erectus* skeleton, the Nariokotome boy, has been hailed as one of the greatest discoveries in palaeoanthropology

(Walker and Leakey 1993). In Europe, important recent finds have been a human tibia from Boxgrove, England (Roberts *et al.* 1994), thought to belong to *H. heidelbergensis*, and the spectacular collection of fossils from Atapuerca in Spain that may include a new species, *H. antecessor* (Arsuaga *et al.* 1993; Bermúdez de Castro *et al.* 1997).

In spite of these discoveries, however, the time when human evolution was established by fossil remains alone has clearly passed: the genetics revolution has meant that we can learn a great deal about our evolutionary past from the laboratory by analysing and comparing samples of DNA from modern humans (Waddell and Penny 1996). Moreover, the significant advances in dating methods during the past decade (such as ESR and Ar/Ar methods) has meant that many human fossils that had long been known, but which either had not been dated, or could be dated only on circumstantial grounds, are now well dated and have caused not a few surprises (Grün and Stringer 1991).

These fossil finds, research in genetics and applications of new dating techniques have helped formulate a more informed view about our evolutionary past. The pattern of human evolution has certainly become clearer, although many details remain to be put in place and major controversies about process continue, notably that about the origin of modern humans.

Comparisons of the DNA of modern humans and chimpanzees have shown that we shared a common ancestor between 5–6 mya (Jones *et al.* 1992). This ancestor most likely lived in Africa as that is where the first fossil remains on direct line to modern humans have been found. But these do not appear until 4.5 mya, and we currently lack any fossil traces, let alone artefacts, of the 6-million-year-old common ancestor to modern humans and chimpanzees.

This is unfortunate, to say the least, as the last common ancestor between ourselves and the great apes is a key species in our evolutionary history. Those aspects of our anatomy, behaviour and cognition that are unique to our species evolved, by definition, after the time of this common ancestor. To understand those developments, we need information about this species. Paradoxically, however, we may be able to ascertain much more about this species than we can about those that come later and have left fossil remains and stone tools. This is because the common ancestor was indeed an ancestor not only to us today, but also to the two species of chimpanzees, *Pan troglodytes* (common chimpanzee) and *Pan piniscus* (the bonobo). Consequently, we can identify features that both we and the chimpanzees share and suggest that these are most likely to have been present in this common ancestor (Byrne 1995: 9–30).

This is the task that Richard Byrne undertakes in his contribution with regard to creative thought. He stresses the value of the comparative method, but also the difficulty of making any progress on the nature of animal thinking. He draws a distinction between 'mundane thought' and 'creative thought'. While he suggests that this distinction is more a matter of behavioural product rather than underlying mechanism, he explains how – paradoxically – it is the latter that indicates the existence of animal thinking at all, since the products of mundane thought are unlikely to be distinguished from thought-free behaviour. Byrne's arguments have many points of contact with those of Ian Hodder in Chapter 4 of this book. Like Hodder, Byrne is reluctant to characterise creative thinking as something discrete and distinct from thought in general, and he too emphasises the importance of understanding situations rather than simply solving problems. For Byrne, creative thinking is part of thought in general, and hence he examines whether modern chimpanzees show evidence for this by exploring how they anticipate tool requirements, plan deception, engage in inter-community violence and control predation. From this evidence, he suggests that the first 'glimmerings of creative thinking' had indeed appeared by the time of the common ancestor. This chapter by Byrne, and indeed his other recent work on the behaviour and thought of the African apes (e.g. Byrne 1995; Byrne and Byrne 1991; Byrne and Whiten 1992), is of immense value to Palaeolithic archaeologists, as they help us to formulate what are the relevant questions to ask about the archaeological record of the first hominids.

AUSTRALOPITHECINES AND EARLY *HOMO*

That record begins 4.5 mya with *A. ramidus*, and extends to *c.*1.8 mya, after which large-brained hominids with modern bipedal, striding gaits and a complex technology had evolved. A hominid dispersion from Africa had either occurred, or was soon to do so (Larick and Ciochan 1996). The intervening period – that of the early hominids – is a critical phase for human evolution as it is when some of the major behavioural distinctions between us and chimpanzees arose, notably bipedalism, the manufacture of stone tools, and a substantial contribution of meat in the diet.

During this period, there was a considerable diversity of hominid species, which fall into two main groups – the australopithecines and early *Homo* (Jones *et al.* 1992). The fomer include *A. afarensis*, made famous by the find of 'Lucy' by Johanson in 1974, and *A. boisei*, first discovered by Mary Leakey at Olduvai Gorge in 1954. Such was the diversity of species overlapping at any one particular point in time that it is not clear which species – or group of species – is responsible for the butchered bones and stone artefacts of the archaeological record (Figure 6.2) (Susman 1991; Wood 1997). There can be little doubt, however, that the early members of the *Homo* genus, *H. habilis* and *H. rudolfensis*, were responsible for many of the stone artefacts and butchered animals found in the archaeological record, simply due to the pervasiveness of tool making among all later members of *Homo*.

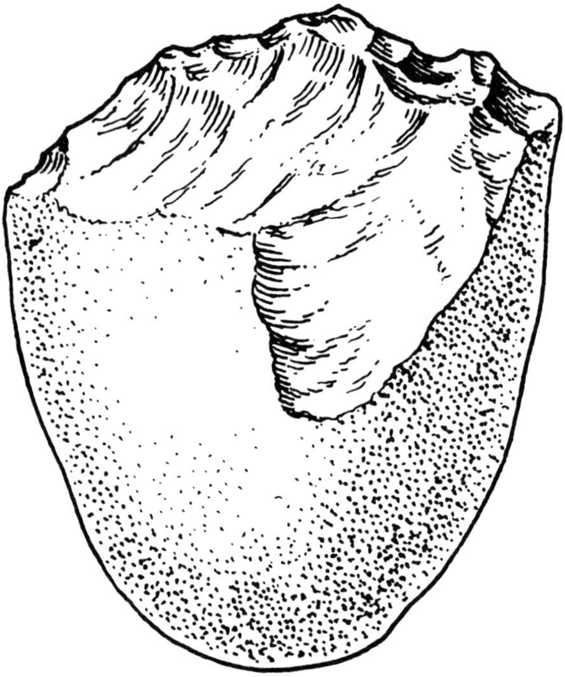

Figure 6.2 An Oldowan chopper from layer I, Olduvai Gorge

Source: Reproduced from F. Bordes, *The Old Stone Age*, 1968, by kind permission of Weidenfeld and Nicolson

The lifestyles of the first members of *Homo* have been subject to intense debate for the last two decades. Models for their behaviour have varied from being essentially modern in form, with home-bases, division of labour and food sharing, to being one of marginal scavenging with social patterns more similar to those of living apes than humans (e.g. Binford 1981; Isaac 1978; Potts 1988). Similarly, the tools of the Oldowan culture have been seen as a technological breakthrough, as little different from the tools made and used by apes, or as something between these two extremes (e.g. Mithen 1996: 96–98; Toth 1985; Wynn and McGrew 1989).

While these controversies continue, there is no doubt that the first *Homo* had a brain significantly larger than that of the australopithecines, reaching about 50 per cent of that of modern humans, was a habitual maker and user of stone tools, and included a substantial amount of meat within his diet (Schick and Toth 1993). Do these features imply abilities at creative thinking? This is the question that Mark Lake addresses in his contribution to this volume.

Lake adopts a different approach from Byrne by beginning with an assumption that creative thinking is found in humans but not in primates.

Consequently, he asks whether this is an attribute of the *Homo* genus, in which case we should find evidence in the behaviour of the first members of that genus, or whether it is something that is just restricted to the final member(s) of that genus. His task is perhaps more challenging than Byrne's, for Lake cannot observe the behaviour of these hominids – he must work with the material residues of behaviour. And those residues may have taken many thousands of years to accumulate and have been transformed by a range of site-formation processes. Nevertheless, Lake shows how the concepts about creativity that Boden discusses in her chapter – those about the exploration of conceptual spaces, and of historical and psychological creativity – are useful for asking questions about the archaeological record of early *Homo*. He examines two issues – the transport of stone and bones across the landscape, and the making of Oldowan artefacts. Lake finds that the data pertaining to the former of these pose 'insurmountable obstacles' in assessing the degree of creative thought, ultimately due to the poor time resolution of archaeological finds, and the constraints on developing a true landscape archaeology. But the situation regarding Oldowan technology seems more favourable towards assessing creative thinking in early *Homo*.

One of the early *Homo* species identified in East Africa, *H. ergaster*, is thought to have been a direct ancestor to *H. erectus* and to have been the first hominid species to move out of Africa (Larick and Ciochan 1996). The record of Eurasian colonisation is notoriously difficult to assess. The last few years have seen claims of *c*.2-million-year-old artefacts from Pakistan (Dennell *et al.* 1988) – which may not be artefacts, or may not be 2 million years old, may be neither of these, or both; the redating of *H. erectus* in Java to 1.8 million years old (Swisher *et al.* 1994) – but which may be substantially younger; and claims for *H. ergaster* in China (Wanpo *et al.* 1995), the evidence for which, however, may not be *Homo* remains at all (Culotta 1995). In Europe, an emerging consensus arose in the early 1990s that there was no occupation prior to 500,000 years ago (Roebroeks and van Kolfschoten 1994), which was then challenged by hominid remains from Atapuerca in Spain more than 750,000 years old (Bermúdez de Castro *et al.* 1997; Parés and Pérez González 1995).

If the date of *Homo* dispersal from Africa is unclear, then so too are how many species of *Homo* existed after 2 mya and their phylogenetic relationships. Some believe that *H. ergaster* should be included within *H. erectus*, while others think that *H. erectus* should be designated as a species evolving in Asia rather than Africa. The new finds from Atapuerca have been claimed to be *H. antecessor* (Bermúdez de Castro *et al.* 1997), the first species to leave Africa, which was ancestral to *H. heidelbergensis* in Europe and *archaic H. sapiens* in Africa. But one can question whether the Atapuercan finds are sufficiently different to constitute a new species; perhaps they should be subsumed within *H. heidelbergensis*.

Now while this brief summary may suggest that we know very little about this period of human evolution, this is in fact far from the case. It is clear that *Homo* did leave Africa soon after 2 million years ago, and that there was a considerable diversity of hominid species. The remarkable find of the Nariokotome boy has given excellent insights into the anatomical adaptations of *H. ergaster*, while it is now clear that modern brain size appeared between 600,000 and 150,000 years ago (Ruff *et al.* 1997). As regards past behaviour, many inferences can be drawn from the fossil remains. The archaeological record is also becoming increasingly well known and understood, particularly from excellently preserved sites such as Boxgrove (Roberts 1986), and ambitious programmes of field work such as at Olorgesailie (Potts 1989).

Perhaps the most notable feature of this archaeological record is the appearance of Acheulian technology at *c.*1.4 mya (Asfaw *et al.* 1992), including handaxes made by the bifacial flaking of nodules of stone or large flakes. This technology shows a substantial advance over that of the Oldowan, as for the first time we find the imposition of form: the distinctive triangle or ovate shape of the handaxe.

Handaxes (Figure 6.3) are one of the great enigmas of our prehistory. As Thomas Wynn (1995) explained, they are completely different from any items of modern culture. On the one hand they seem to encapsulate the absence of creativity – the repetition of the same technology, and often the same form, over many thousands of years and across vast expanses of space, probably made by several different species (Wynn and Tierson 1990). Yet a moment's reflection suggests that handaxes may be indicating the precise opposite – a very creative mind in terms of creativity as interpretation and understanding. For while the hominids making them appear to have been aiming for very similar results, they began with nodules of stone of different shapes and sizes and needed to undertake a different sequence of knapping actions to have arrived at this result. We seem to be seeing a creative manipulation of a limited set of knapping actions and technological ideas, which reminds us of the relationship between constraint and creativity introduced by Boden and to which later chapters will return.

THE NEANDERTHALS AND THE ORIGINS OF MODERN HUMANS

The chapter following that by Lake deals with another great enigma of prehistory: the Neanderthals. *H. neanderthalensis* is now thought to be a descendant of *H. heidelbergensis* and was a species restricted in range to Europe and western Asia (Stringer and Gamble 1993). Physiologically, Neanderthals were adapted for cold climates and the variation in their geographical distribution during the Pleistocene most likely reflects the changes in climate: as it became warmer, they moved into more northerly latitudes; as it became colder they moved south.

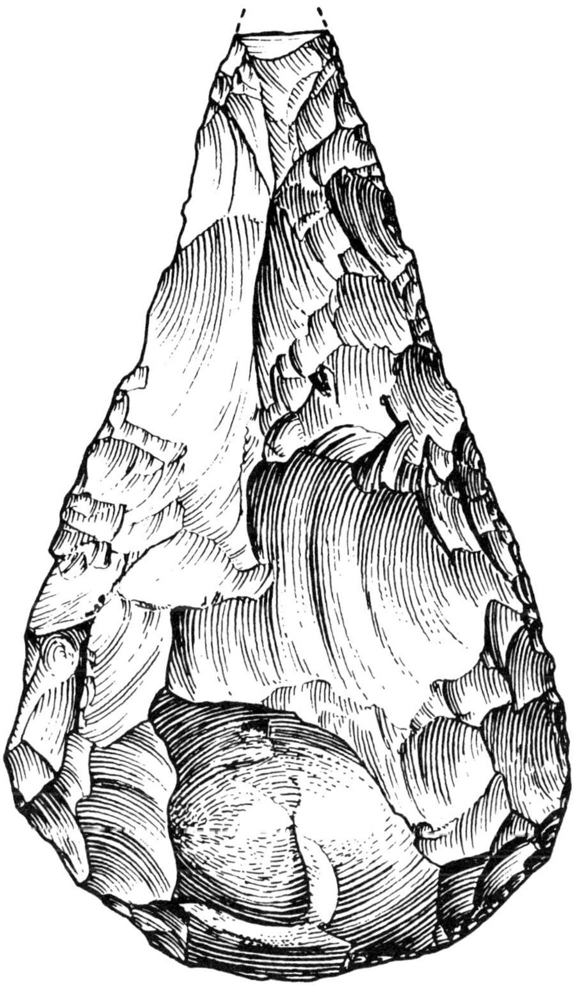

Figure 6.3 Handaxe from Swanscombe, England
Source: Reproduced from F. Bordes, *The Old Stone Age*, 1968, by kind permission of Weidenfeld and Nicolson

Ever since Neanderthals were first discovered in the nineteenth century, two issues have been central in the debates about them: what is their relationship to modern humans, and what was the nature of their minds? The last few years have seen a flurry of books about the Neanderthals, with particularly important works by Steven Kuhn (1995) and Mary Stiner (1995) on Neanderthal behaviour in west central Italy, and a comprehensive synthesis of the data from western Europe in general by Paul Mellars (1996; see also Stringer and Gamble 1993; Trinkaus and Shipman 1993). The nature of the Neanderthal mind has been discussed in much of this literature, particularly with regard to their linguistic and symbolic abilities (see also papers in Mellars and Gibson 1996; Mithen 1996; Noble and Davidson 1996). But whether we should think of Neanderthals as creative thinkers and doers has not been adequately addressed.

This is the challenge taken up by Mary Stiner and Steven Kuhn in this volume. As with Lake, they face up to the problems posed by working with data from the archaeological record, as opposed to direct observations of behaviour. How can creative acts come to be visible in the archaeological record? Many factors appear to militate against this – the processes of preservation, those of discovery, and the fact that so much of behaviour may leave no archaeological trace. It is indeed particularly frustrating that it is within the social interactions of living primates that the greatest evidence appears for complex cognitive skills – a Machiavellian intelligence (Byrne and Whiten 1988) – when this domain of behaviour is one of the most difficult for Palaeolithic archaeologists to reconstruct.

Like Byrne, Stiner and Kuhn find themselves with a paradox. Byrne suggested that finding traces of creative thinking is the best way to identify thinking of any type among living primates; Stiner and Kuhn argue that it is only due to the apparent lack of creative thinking in the majority of Neanderthals, and their seeming compulsion for imitation, that we may be able to recognise evidence of creative thinking by the few. Had it not been for the high degrees of imitation, then any innovations are not likely to become sufficiently widespread to leave an archaeological trace.

Stiner and Kuhn find more paradoxes in the archaeological record of the Neanderthals. Although there is circumstantial evidence that Neanderthals must have displayed considerable flexibility in their foraging and social behaviour, their weapons technology shows remarkable consistency across time and space. As Stiner and Kuhn demonstrate, Neanderthals' stone weapon tips – Levallois and Mousterian points (Figure 6.4) – are universally made to the same plan, even though the specific technical acts in producing these appear variable. In this sense, we appear to see no evidence of creative thinking at all in their design of hunting weapons. Whether this indicates an equal absence of creativity in other domains of life remains unclear.

The fundamental reason that Stiner and Kuhn offer regarding the lack of innovation and variability in weapons is that Neanderthal technology was not embedded within other domains of behaviour, such as social interaction and foraging, in the same manner as is the technology of modern humans. In this regard, the conceptual space of technology appears to have been extremely narrow and constrained – quite unlike technology when we think of it today. Stiner and Kuhn arrived at this idea from a detailed examination of the evidence from one specific region of Europe (Kuhn 1995: 174). I independently arrived at a similar conclusion from a theoretical perspective, beginning with general ideas about how the modern mind may have evolved, prior to considering the archaeological data

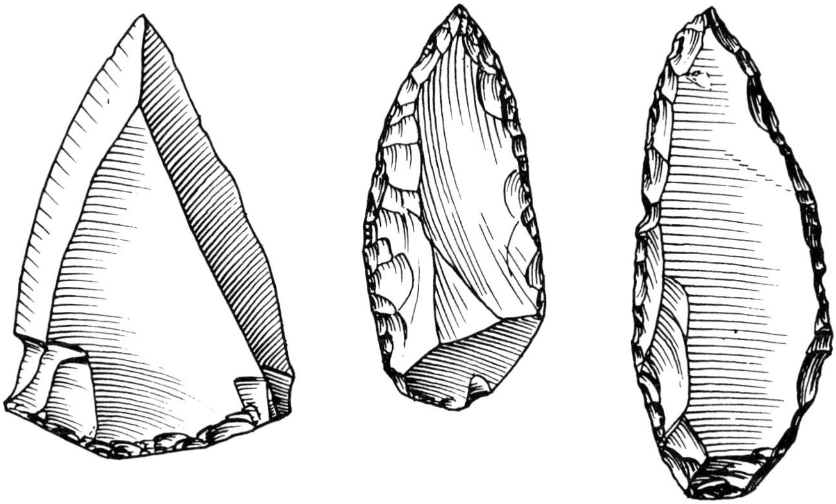

Figure 6.4 (a) Levallois point, from Houppeville, Normandy; (b, c) Mousterian points, from Combe Grenal, Dordogne, layer 29
Source: Reproduced from F. Bordes, *The Old Stone Age*, 1968, by kind permission of Weidenfeld and Nicolson

(Mithen 1996). In this light, therefore, the 'domain-specific' mentality of Neanderthals is emerging as a particularly robust idea, and one that appears fundamentally linked to our understanding of human creativity and valuable to those concerned either with interpreting the nitty-gritty details of the archaeological record or with general theories of cognitive evolution.

It is most likely that Neanderthals were not directly related to modern humans, but consituted an evolutionary dead end (Stringer and McKie 1996). The origins of modern humans has been *the* major debate in palaeoanthropology during the last decade, with ardent advocates of two opposing theories: the out-of-Africa theory, which posits a single, recent origin in Africa, and a multi-regional theory, which suggests multiple origins throughout the old world (Aiello 1993 provides a good summary). This is not the place to rehearse the arguments for each side, which have been so thoroughly discussed elsewhere (for two volumes with opposing views and which cover the history of this debate seen from different perspectives see Stringer and McKie 1996 and Wolpoff and Caspari 1997). My view is firmly on the side of the out-of-Africa argument, with some inter-breeding between a dispersing modern human population and existing archaic humans in Europe and Asia, including Neanderthals.

The evidence from genetics suggests that modern humans had appeared by 250,000 years ago, while the first fossil traces are found by 100,000 years ago in both south Africa and western Asia. By 50,000 years ago, it seems that modern humans had dispersed to many parts of the Old World, including Australia. In this light, it is perhaps remarkable that the archaeological record between 250,000 and 50,000 shows such limited variability. There appear no general correlations between material culture and hominid species (but see Lieberman and Shea 1994 for subtle differences between Neanderthals and early modern humans in the Near East, and Foley and Lahr 1997 for an opposing view). It is indeed the continuities in the archaeological record in many parts of the world throughout the Upper Pleistocene that has been used as one of the supports for the multi-regional theory of modern human origins (Frayer *et al.* 1993).

It is only in western Europe that there appears to be a reasonably clear association between the characteristics of the archaeological record and the associated hominid species (Mellars 1989). The modern humans who spread into Europe after 40,000 years ago appear to have possessed a distinctly different suite of material culture items, including a blade-based stone tool technology, artefacts made from bone and antler and items of body decoration. But even here the associations are rather blurred, as some Neanderthals seem to be associated with a similar material culture, which may have been made in imitation of modern humans, or scavenged from their occupation sites (Harrold 1989).

Yet this date of 40,000 years ago falls within a window of change in the archaeological record throughout the world (Mithen 1996). This stretches between 60,000 years ago – when people most likely reached Australia (Allen 1994) – and 25,000 years ago, after which there is clear evidence for art in Africa, Asia and Europe, and people had colonised arid landscapes which earlier types of humans had never penetrated (Gamble 1993).

Many archaeologists believe that this window of change represents a fundamental evolutionary event in the making of modern humans, and one that lies at the ultimate root of the modern world (for an opposing view see Foley and Lahr 1997). It is normally described using the European terminology of the Middle/Upper Palaeolithic transition and is epitomised by the appearance of art, especially that in Europe, such as the paintings in Chauvet cave (Chauvet *et al.* 1996). It is only after this window of change that we find artefacts and behaviours that are the usual subject for studies of human creativity – objects of art and technological inventions (Figure 6.5).

The final chapter in this section, by myself, discusses the Middle/Upper Palaeolithic transition – an event that has been characterised as the creative explosion (Pfeiffer 1982). This contribution does not deny that creativity is a term that can be validly applied to aspects of behaviour of the common ancestor, the first *Homo*, or of the Neanderthals. But creative thinking and action after 30,000 years ago appears to be of a different order of magnitude from that which went before. In Boden's terms, the conceptual spaces and the extent of exploration appear to have

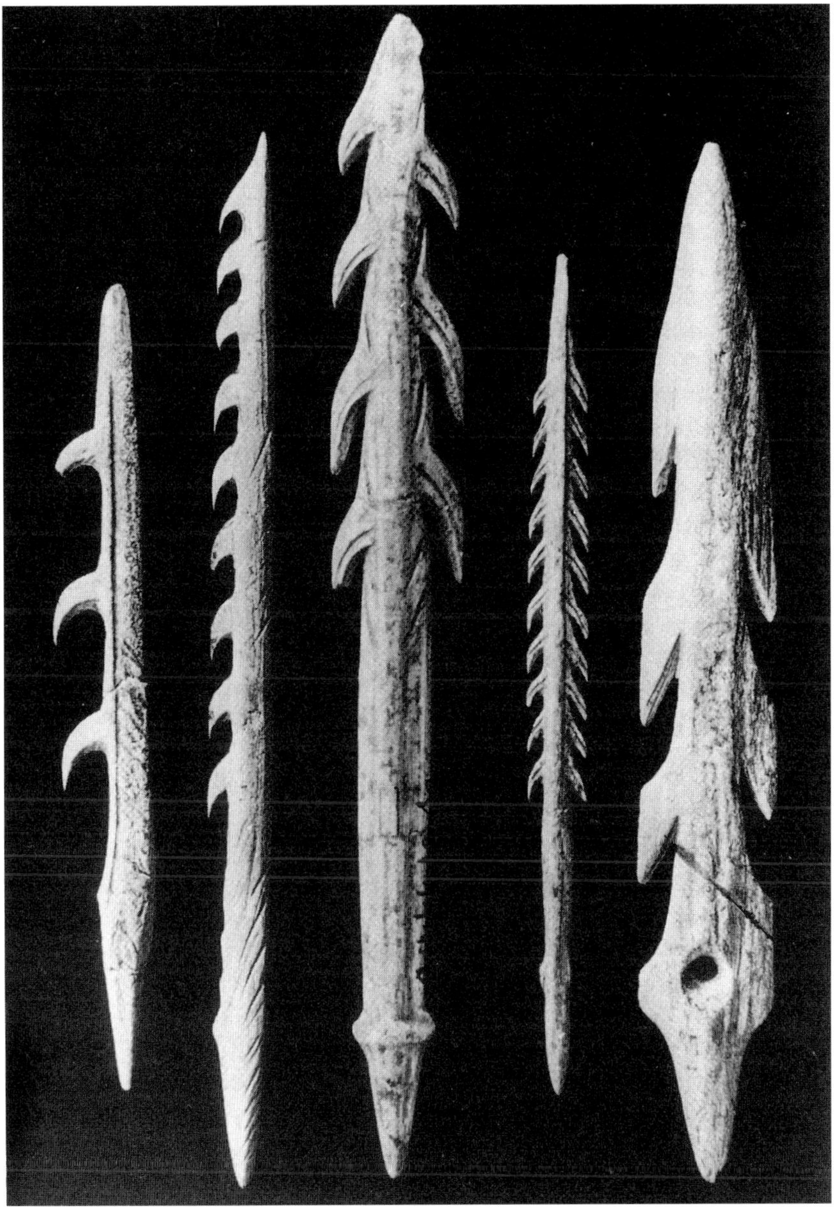

Figure 6.5 Harpoons of the later European Upper Palaeolithic
Source: Reproduced from F. Bordes, *The Old Stone Age*, 1968, by kind permission of Weidenfeld and Nicolson

become dramatically transformed. Quite how this happened is discussed by examining three foundations for modern creativity: the possession of a theory of mind, language and a material culture that functions as an extension of the mind. A consequence of the new found creativity was the use of material culture to actively structure interactions with the social and natural worlds, resulting in the eventual competitive exclusion of all other human types, leaving *H. sapiens sapiens* as the only surviving member of the *Homo* genus.

By means of this extended editorial introduction I have explained how the four chapters within this section examine the evolution of creativity, from the common ancestor of 6 million years ago to those humans whom we identify as anatomically, behaviourally and cognitively modern. The chapters address only a fraction of the issues in the archaeological and fossil record of human evolution that might help to illuminate the phenomenon of creativity. I hope they will stimulate new research to examine further issues from a similar perspective, such as colonisation and Acheulian technology, as well as furthering our understanding of cognitive evolution and the nature of human creativity.

REFERENCES

Aiello, L. (1993) 'The fossil evidence for modern human origins in Africa: a revised view.' *American Anthropologist* 95: 73–96.

Aitchison, J. (1996) *The Seeds of Speech, Language Origin and Evolution*. Cambridge: Cambridge University Press.

Allen, J. (1994) 'Radiocarbon determinations, luminescence dates and Australian archaeology.' *Antiquity* 68: 339–343.

Arsuaga, J.-L., Martínez, I., Gracia, A., Carretero, J.-M. and Carbonell, E. (1993) 'Three new human skulls from the Sima de los Huesos Middle Pleistocene site in Sierra de Atapuerca, Spain.' *Nature* 362: 534–537.

Asfaw, B., Beyene, Y., Suwa, G., Walter, R.C., White, T., Wolde-Gabriel, G. and Yemane, T. (1992) 'The earliest Acheulean from Konso-Gardula.' *Nature* 360: 732–735.

Barkow, J., Cosmides, L. and Tooby, J. (1992) *The Adapted Mind: Evolutionary Psychology and the Evolution of Culture*. Oxford: Oxford University Press.

Bermúdez de Castro, J.M., Arsuaga, J.-L., Carbonell, E., Rosas, A., Martínez, I. and Mosquera, M. (1997) 'A hominid from the Lower Pleistocene of Atapuerca, Spain: Possible ancestor to Neanderthals and Modern humans.' *Science* 276: 1392–1395.

Bickerton, D. (1996) *Language and Human Behaviour*. London: UCL Press.

Binford, L.R. (1981) *Bones: Ancient Men and Modern Myths*. New York: Academic Press.

Bordes, F. (1968) *The Old Stone Age*. London: Weidenfeld & Nicolson.

Byrne, R.W. (1995) *The Thinking Ape: Evolutionary Origins of Intelligence*. Oxford: Oxford University Press.

Byrne, R.W. and Byrne, J.M.E. (1991) 'Hand preferences in the skilled gathering tasks of mountain gorillas (*Gorilla g. beringei*).' *Cortex* 27: 521–546.

Byrne, R.W. and Whiten, A. (eds) (1988) *Machiavellian Intelligence: Social Expertise and the Evolution of Intellect in Monkeys, Apes and Humans.* Oxford: Clarendon Press.
Byrne, R.W. and Whiten, A. (1992) 'Cognitive evolution in primates: evidence from tactical deception.' *Man* (n.s.) 27: 609–627.
Chauvet, J.-M., Deschamps, E.B. and Hillaire, C. (1996) *Chauvet Cave: The Discovery of the World's Oldest Paintings.* London: Thames & Hudson.
Corballis, M.C. (1992) 'On the evolution of language and generativity.' *Cognition* 44: 197–226.
Corballis, M.C. (ed.) (in press) *The Evolution of the Hominid Mind.* Oxford: Oxford University Press.
Culotta, E. (1995) 'Asian hominids grow older.' *Science* 270: 1116–1117.
Cummins, D. and Allen, C. (eds) (in press) *Evolution of the Mind.* Cambridge, MA: MIT Press.
Dennell, R.W., Rendell, H. and Hailwood, E. (1988) 'Early tool making in Asia: two million year old artefacts in Pakistan.' *Antiquity* 62: 98–106.
Dunbar, R.I.M. (1996) *Grooming, Gossip and the Evolution of Language.* London: Faber & Faber.
Foley, R. (1987) *Another Unique Species.* Harlow: Longman.
Foley, R. and Lahr, M. (1997) 'Mode 3 technologies and the evolution of modern humans.' *Cambridge Archaeological Journal* 7: 3–36.
Frayer, D.W., Wolpoff, M.H., Thorner, A.G., Smith, F.H. and Pope, G.G. (1993) 'Theories of modern human origins: the paleontological test.' *American Anthropologist* 95: 14–50.
Gamble, C. (1993) *Timewalkers: The Prehistory of Global Colonization.* Stroud: Alan Sutton.
Grün, R. and Stringer, C. (1991) 'Electron spin resonance dating and the evolution of modern humans.' *Archaeometry* 33: 153–199.
Harrold, F. (1989) 'Mousterian, Chatelperronian and early Aurignacian in western Europe: continuity or discontinuity?' In P. Mellars and C. Stringer (eds) *The Human Revolution*, pp. 677–713. Edinburgh: Edinburgh University Press.
Isaac, G. (1978) 'The food-sharing behaviour of proto-human hominids.' *Scientific American* 238 (April): 90–108.
Jones, J.S., Martin, R. and Pilbeam. D. (eds) (1992) *The Cambridge Encyclopedia of Human Evolution.* Cambridge: Cambridge University Press.
Kuhn, S. (1995) *Mousterian Lithic Technology.* Princeton, NJ: Princeton University Press.
Larick, R. and Ciochan, R.L. (1996) 'The African emergence and early African dispersal of the genus *Homo*.' *American Scientist* 84: 538–551.
Leakey, M., Feibel, C.S., McDougall, I. and Walker, A. (1995) 'New four million year old hominid species from Kanapoi and Allia Bay, Kenya.' *Nature* 376: 565–571.
Lieberman, D. and Shea, J. (1994) 'Behavioural differences between archaic and modern humans in the Levantine Mousterian.' *American Anthropologist* 96: 330–332.
Mellars, P. (1989) 'Major issues in the emergence of modern humans.' *Current Anthropology* 30: 349–385.
Mellars, P. (1991) 'Cognitive changes and the emergence of modern humans.' *Cambridge Archaeological Journal* 1: 63–76.

Mellars, P. (1996) *The Neanderthal Legacy*. Princeton, NJ: Princeton University Press.

Mellars, P. and Gibson, K. (eds) (1996) *Modelling the Early Human Mind*. Cambridge: McDonald Institute for Archaeological Research.

Mithen, S. (1993) 'From domain-specific to generalised intelligence: a cognitive interpretation of the Middle/Upper Palaeolithic transition.' In C. Renfrew and E. Zubrow (eds) *The Ancient Mind*. pp. 29–39. Cambridge: Cambridge University Press.

Mithen. S. (1996) *The Prehistory of the Mind: A Search for the Origins of Art, Science and Religion*. New York and London: Thames & Hudson.

Noble, W. and Davidson, I. (1996) *Human Evolution. Language and Mind*, Cambridge: Cambridge University Press.

Parés, J.M. and Pérez-González, A. (1995) 'Paleomagnetic age for hominid fossils of Atapuerca archaeological site.' *Science* 269: 830–832.

Pfeiffer, J. (1982) *The Creative Explosion*. New York: Harper & Row.

Potts, R. (1988) *Early Hominid Activities at Olduvai Gorge*. New York: Aldine de Gruyter.

Potts, R. (1989) 'Olorgesailie: new excavations and findings in Early and Middle Pleistocene contexts, southern Kenya rift valley.' *Journal of Human Evolution* 18: 269–276.

Roberts, M.B. (1986) 'Excavation of the Lower Palaeolithic site at Amey's Eartham Pit, Boxgrove, West Sussex: a preliminary report.' *Proceedings of the Prehistoric Society* 52: 215–246.

Roberts, M.B., Stringer, C.B. and Parfitt, S.A. (1994) 'A hominid tibia from Middle Pleistocene sediments at Boxgrove, U.K.' *Nature* 369: 311–313.

Roebroeks, W. and van Kolfschoten, T. (1994) 'The earliest occupation of Europe: a short chronology.' *Antiquity* 68: 489–503.

Ruff, C.B., Trinkaus, E. and Holliday, T.W. (1997) 'Body mass and ecaphization in Pleistocene *Homo*.' *Nature* 387: 173–176.

Schick, K. and Toth, N. (1993) *Making Silent Stones Speak: Human Evolution and the Dawn of Technology*. New York: Simon & Schuster.

Stiner, M. (1995) *Honor Among Thieves: A Zooarchaeological Study of Neanderthal Ecology*. Princeton, NJ: Princeton University Press.

Stringer, C. and Gamble, C. (1993) *In Search of the Neanderthals*. London: Thames & Hudson.

Stringer, C. and McKie, R. (1996) *African Exodus: The Origins of Modern Humanity*. London: Jonathan Cape.

Susman, R.L. (1991) 'Who made the Oldowan tools? Fossil evidence for tool behaviour in Plio-Pleistocene hominids.' *Journal of Anthropological Research* 47: 129–151.

Swisher, C.C. III, Curtis, G.H., Jacob, T., Getty, A.G., Suprijo, A. and Widiasmoro, (1994) 'Age of the earliest known hominids in Java, Indonesia.' *Science* 263: 1118–1121.

Toth, N. (1985) 'The Oldowan reassessed: a close look at early stone artefacts.' *Journal of Archaeological Science* 12: 101–120.

Trinkaus, E. and Shipman, P. (1993) *The Neanderthals*, New York: A.A. Knopf.

Waddell, P.J. and Penny, D. (1996) 'Evolutionary trees of apes and humans from DNA sequences.' In A. Lock and C.H. Peters (eds) *Handbook of Human Symbolic Evolution*, pp. 53–73. Oxford: Oxford University Press.

Walker, A. and Leakey, R. (1993) *The Nariokotome Homo erectus Skeleton*. Berlin: Springer Verlag.

Wanpo, H., Ciochan, R., Yumin, G., Larick, R., Qiren, F., Schwarcz, H., Yonge, C., de Vos, J. and Rink, W. (1995) 'Early *Homo* and associated artefacts from China.' *Nature* 378: 275–278.

Whallon, R. (1989) 'Elements of culture change in the Later Palaeolithic.' In P. Mellars and C. Stringer (eds) *The Human Revolution*, pp. 433–454. Edinburgh: Edinburgh University Press.

White, T.D., Suwa, G. and Asfaw, B. (1994) '*Australopithecus ramidus*, a new species of early hominid from Aramis, Ethiopia.' *Nature* 371: 306–312.

Wolpoff, M. and Caspari, R. (1997) *Race and Human Evolution*. New York: Simon & Schuster.

Wood, B. (1997) 'The oldest whodunnit in the world.' *Nature* 385: 292–293.

Wynn, T. (1995) 'Handaxe enigmas.' *World Archaeology* 27: 10–23.

Wynn, T. and McGrew, W.C. (1989) 'An ape's view of the Oldowan.' *Man* 24: 383–398.

Wynn, T. and Tierson, F. (1990) 'Regional comparison of the shapes of later Acheulean handaxes.' *American Anthropologist* 92: 73–84.

CHAPTER SEVEN

THE EARLY EVOLUTION OF CREATIVE THINKING

Evidence from monkeys and apes

RICHARD BYRNE

In everyday parlance, 'creative thought' is an exalted category, far above run-of-the-mill 'thinking' and 'comprehending', and associated instead with major innovation and discovery. People like Isaac Newton, Bertrand Russell and John Donne think creatively; the rest of us just think. In psychology, the term is viewed a little differently.

The impact of Piaget's work has led to an appreciation that even the fumblings of a young child can embody creative discovery – but at the individual and personal level. This distinction is captured by Boden's term 'psychological-creativity' versus the ordinary, exalted sense of historical-creativity (see Chapter 3). In terms of the underlying cognitive mechanism, an organism capable of psychological-creativity is in principle also capable of historical-creativity: the distinction depends only on whether the personal discovery is known already to the community at large. In early childhood, there is so much to assimilate and understand that psychological-creativity becomes almost routine. Nevertheless, the requirement of personal discovery and innovation means that most of us, most of the time, do not consider that we are busy in creative thought.

Instead, most modern humans spend much of each day thinking in more mundane ways: planning where to go, what to do, what to eat, what to say to whom, and so on. These everyday activities are, however, exceedingly hard to simulate mechanically, as many stymied attempts in artificial intelligence testify. Thanks to the AI enterprise, realisation has grown that everyday thinking and planning are activities that, despite their familiarity, involve manipulations of mentally represented information that are complex and ill understood. With this new perspective, the gap between mundane and creative thought is considerably narrowed. Great thinkers may have more control than most people over their ability to re-conceptualise a problem space to allow novel insights, and may be prone to form their novel associations at a higher level of abstraction. But the fact that ordinary people are able to think and plan – at all – almost guarantees that every now and then new connections or new ways of looking at a problem will be found. These insights are small ones, but still require novel facts to be computed: the realisation that a new short cut can be taken to a familiar destination, that vegetarian food helps calorie reduction, that Susan may not like you despite her ostensibly friendly manner. Indeed, *any* organism that could do the everyday planning tasks that people so often take for granted could probably also, on occasion, make a creative leap. Thus, as with the psychological- vs. historical-creativity distinction, the difference between mundane thought and creative thought is more a matter of product than underlying mechanism. Moreover, *understanding* an existing social situation or how a simple mechanism works are themselves activities that require subtle and ill-understood processes of thought, just as much as does creative thinking (and see Hodder, Chapter 4). The term 'mundane thought' is almost a contradiction in terms: thought is not a mundane activity.[1]

This prologue serves to argue that in order to understand the origins of the potential for creative thought, we need only ask about the origins of *thinking*, of the kind carried out daily by most people. This cuts the problem down to a more manageable size. Nevertheless, the ephemeral nature of any sort of thinking poses a severe problem for historical enquiry, let alone for probing the evolutionary origins of the ability. We cannot speak to our ancestors, to ask them of their thoughts; we cannot observe their behaviour, in order to deduce their thinking; and their behaviour mostly left no tangible remains to dig up. How might we proceed?

THE COMPARATIVE METHOD

When we ask about the evolutionary history of any human cognitive process, we ideally need to use a comparative approach, getting behavioural evidence from modern primates or more distant relatives. Where a group of species are all more closely related to each other than they are to any other species (i.e. they all share a single common ancestor), the group is called a *clade*. If a behavioural trait is found in all members of a clade, but not in other less closely related species, the inception of the trait can safely be attributed to the (inferred) ancestor of the clade. This method is powerful, because it relies on evidence only from extant species, in which behaviour can be studied directly. It thus circumvents the problems unavoidable with the use of the fossil record. Quite apart from the sparseness of any such record, the indirect deduction of behaviour from its concrete remains suffers from uncertainty both about exactly what behaviour resulted in the remains, and also about which of the various possible fossil species was responsible. Importantly, in using a comparative approach, the existence and dating of ancestor species is inferred from the *same* species, the ones whose behaviour is studied. Usually nowadays this is done by using molecular taxonomy. While the dates so derived are sometimes controversial and are always bracketed with error terms, this method avoids the problems inherent in relating fossils to modern forms. In the perhaps apocryphal reported words of Vincent Sarich, 'I know my molecules had ancestors, you have to hope your fossils had descendants.' When molecular taxonomy and dating methods were first introduced, they generated a storm of criticism, with frequent assertions that the conclusions must be wrong because they contradicted the fossil record. In retrospect, it would appear that the phantom was the solidity of the fossil record; for instance, the recency of the chimpanzee's separation from the human line was a particular point of dispute, yet even now we lack a single fossil of a chimpanzee's direct ancestor.

The only serious limitation of the comparative method is that its reliability is a function of the number of related species that are compared. Unfortunately, the closest surviving relatives of humans – the great apes – are few in number; and, of course, the method cannot be applied to a single surviving species, the case for the last 5 million years of hominid evolution. For that period, only careful archaeology can illuminate the conceptual abilities of our ancestors. This chapter will delve into the earlier origins of cognition, using the comparative method. The aim will be to discover whether our ancestors, over that period of human ancestry that we share with non-human primates (from 65 to 5 mya), were able to think, and if so at what point this arose. While firm claims would be premature, the balance of evidence supports the view that this occurred before the first hominids. This history also permits some informed speculation as to what selective pressures led to this novel adaptation, and what anatomical changes underpinned its occurrence.

EVALUATING EVIDENCE OF ANIMAL THINKING

Until very recent years, the idea that animals – even primates – might engage in *any* sort of thinking was regarded as radical, or more likely sentimental nonsense. Two forces have changed this bleak assessment. The gradual acceptance within psychology of mechanistic theories of thought processes (Miller *et al.* 1960; Newell *et al.* 1958; Newell and Simon 1972) have lessened the mystique of human 'thought' and brought into being the subject of 'cognitive psychology', in which there is no serious doubt that thinking will one day be fully understood. In ethology too there has been a growing realisation that testing predictions derived from taking an 'intentional stance' (Dennett 1983), or a belief that animals think (Griffin 1976, 1984), can lead to surprising confirmations on occasion.

As an example, it had long been taken for granted that the 'broken wing' distraction display that plovers and other species of birds perform when a predator approaches their nest was a hard-wired routine, enacted in an inflexible and rigid way once triggered. But when a researcher (explicitly under the influence of Griffin's thinking) designed experiments to check this, plovers proved to be sensitive to the actions and direction of attention of predators, and to perform the display in a goal-directed way, repeating it in a more favourable location if the predator was not at first distracted (Ristau 1991). This is not to suggest that ethologists have abandoned caution wholesale and now assume that there is thinking in animals until proved otherwise. Instead, wise ethologists now take an agnostic stance (Dawkins 1993), arguing both with those who are sure that their dog thinks all the time (that they may quite possibly be wrong) and with those who cannot imagine that any non-linguistic being could possibly think (that even this heresy is possible).

When an intelligent-seeming behaviour is observed in animals, the first explanation that springs to the minds of ethologists is not that the animal thought it out, but that associative learning mechanisms may have automatically built up programmes of behaviour without any active, computational process occurring in the organism. At first sight, this might seem an unreasonable bias, but it must be realised that unstructured, associative learning is capable of accounting for surprisingly complex behaviour (Dickinson 1980) – even in humans, if we're to be fair. Learning of this sort occurs when events are repeatedly associated in space and time, either as a result of observing their spontaneous co-occurrence, or when a self-generated exploratory action is *reinforced* by some reward occurring just when it is done. Both forms are sometimes referred to as 'conditioning', Pavlovian and operant respectively. Over the years, many remarkably clever-looking animal behaviours have proved to be explicable as a product of these rather simple learning mechanisms.

One of the most striking examples comes from the Japanese monkeys on Kosima Island. This community of monkeys has developed various novel food-processing skills, for instance washing dirty sweet potatoes and extracting wheat from sand by throwing it into water (Nishida 1986). This is often claimed to represent a simple culture: an innovation by a 'genius' or lucky monkey is transmitted socially to new users and eventually across generations. In fact, these behaviours were probably unintentionally reinforced by the people who fed the monkeys – who, quite naturally, would tend to give slightly more generously to monkeys that did any such cute actions (Green 1975). The inception of the habit, rather than being a result of innovation, may also have been caused by chance human reinforcement of monkey actions. If monkeys had learnt from successful foragers by imitation, as was initially thought to be the case, the spread should have been fast. In fact it was very slow, and even after several years many monkeys had not learnt the methods (Galef 1988). Under more restricted conditions of captivity, monkeys discover how to wash dirty food with surprising speed, acquiring this habit after only two hours of exposure to dirty food and water, by means of piecemeal accumulation of actions that produce rewarding consequences, without any sign of quiet, 'thoughtful' contemplation or close observation and imitation of others (Visalberghi and Fragaszy 1990). Associative learning is made more effective, in highly social species like monkeys, by means of an innate tendency to attend closely to the site of a conspecific's successful food procurement, a tendency called *stimulus enhancement* (Spence 1937); in a range of social species, local traditions of behaviour are now known (Roper 1983; Terkel 1994). However, in the absence of good evidence for imitation, most researchers would be reluctant to term a local tradition of behaviour a 'culture' (Tomasello 1990).

Learning can also give the illusion of understanding cause-and-effect. Capuchin monkeys readily learn to use sticks to push peanuts out of a transparent tube: they appear, then, to understand the process. When a 'trap' is made in the tube, into which peanuts can fall and become inaccessible, tool-using monkeys are at first baffled (Visalberghi and Limongelli 1994, 1996). One monkey did learn in which direction it is safe to push the peanut, but by a long process of trial-and-error. And when the trap was then rotated, so that it had no effect, the monkey's behaviour remained the same, fixated on avoiding a problem that humans can see is not there. Evidently monkeys learn to perform specific movements, but do not understand the mechanics of the process that delivers success: the cause-and-effect relations of pushing and the effects of gravity.

Monkey 'culture' and 'tool using' are therefore to some extent a sham, unlike their human counterparts. Nevertheless, the behavioural products are impressive to see, and had convinced many observers that monkeys were capable of understanding cause-and-effect, learning new skills by imitation, and even working out solutions by thinking. These cautionary tales show that it will not be easy to discriminate conclusive evidence of thinking from the appropriate 'null hypothesis' of associative learning. How might we begin? Since we must rely entirely on behaviour to detect thinking in animals, we need some behavioural sign of thinking that is different from that predicted by learning. Most usefully for this purpose, the process of thinking is unique in that it is capable of deriving *solutions that are novel in the experience of the individual, in the absence of sufficient opportunities for learning* to have produced the same results. Since novelty is so important in the identification of thinking, creative thought is actually more likely than mundane thought to have such consequences. Mundane thought would be hard to tell from (thought-free) learning, so in animals oddly enough we are more likely to gain evidence of creative uses of thinking.

Evidence of this kind might come from *experiments*, in which entirely new circumstances are set up for individuals and novel solutions may result, or from spontaneous *observations* of solutions that cannot plausibly have resulted from past experience. The former might seem the safer method, since it does not need to rely on judgements of 'plausibility' – and for animals bred in the laboratory and kept singly this is true. However, for animals living naturally in social groups, it is normally impossible to know the precise history of each individual, and once again plausibility must be invoked. Unfortunately, individually housed laboratory animals of solitary species seldom perform any behaviour of sufficient complexity to suggest that it needed to be thought out, and in the case of highly social species like monkeys and apes there is every reason to think that such housing causes intellectual impoverishment (and cruelty), so most evidence comes from somewhat more natural conditions.

If we accept as an indication of thinking any evidence that an individual is doing something that it was not genetically programmed to do, or that it had not learnt explicitly to do in the past, then there are a number of lines of evidence that suggest thinking in animals. (Ruling out genetical pre-programming and associative learning as alternative explanations is a major undertaking in many cases.) I will argue that, while no single observation is fully convincing on its own, when put together the evidence cannot be parsimoniously explained without allowing the potential for creative thought in some species of primates.

ANTICIPATING TOOL REQUIREMENTS

Perhaps the greatest advantage of being able to think is that one can anticipate problems ahead of time, rather than just react to circumstances. As an everyday example, if we intend to dig the garden, we anticipate needing a fork, realise it is locked in the garden shed, and carry the key with us as we leave the house. Tools give many opportunities for this sort of foresight: do non-human primates show it? Among wild primates, tool use is confined to chimpanzees.[2] Chimpanzees sometimes show anticipation of future requirements for tool use. At one West African site, large round stones are used to crack very hard nuts, using flatter stones as anvils (Boesch and Boesch 1981, 1983). In tropical forests, conveniently sized cobbles for use as hammers are not common, and so chimpanzees regularly pick up and carry a 'hammer stone' when they are going to a nut-bearing tree, sometimes carrying the stone many hundreds of yards. Analysis of their travel routes has shown that they take the minimal-distance path (Boesch and Boesch 1984), so we can be sure that they do not simply wander with the hammer stone until they happen upon a source of nuts. Evidently, they already know what object they will later need, for the specific job of nutcracking, well in advance of the situational context. The same is found in cases where tools are made, not found. At the many sites where chimpanzees use thin probes of plant stems to fish for termites, when there is no source of probes near the termite mound, chimpanzees will prepare the tools in advance (Goodall

1973, 1986). This involves selecting and pulling off a stem of suitable thickness and flexibility, removing side-leaves and often biting the end to give it a clean tip. Several tools may be prepared, then carried to the termite mound. Again, chimpanzees evidently know what they need from a tool, before reaching the site of use. This evidence implies that the *intention* to feed on nuts or termites, in the absence of the food itself, sets a process in motion that causes advance selection and preparation of tools. Such a process would normally, of course, be called thinking: generating a representation of some object, unprompted by stimulus cues from the environment.

In captivity, capuchin monkeys readily use objects as tools, a habit that has enabled a test of whether they too are able to evoke an advance representation of the tool they need for a particular job. Capuchins that had become used to extracting a peanut from a tube – using a particular stick as a probe – were given the task with the stick now removed (Visalberghi and Limongelli 1994, 1996). Instead, they were given a choice of various other objects. One might expect them to attempt the task with the object most similar in features to the familiar probe: thin enough to enter the hole, long enough to emerge at the end, etc. However, unlike chimpanzees, they did not seem to know what an appropriate tool should look like: instead, they tried blocks that were too thick to go in the tube, sticks too short to reach the end, and even on one occasion a piece of flexible chain. This is evidence that capuchins lack a mental representation of what sort of tool is needed. Nevertheless, their rapid learning ability and tendency to explore objects enables them to become proficient tool users when presented with tasks that can be solved by using some available material as a tool. Without the ability to use mental representations, monkeys must remain tied to the immediately present stimulus configuration, unable to compute what it is the situation demands. Mental representation corresponds to 'understanding', in everyday terms. The difference in behaviour of tool-using monkeys and apes thus shows how understanding and thinking are intimately related.

PLANNING DECEPTION

Human social manipulations present rich opportunities for anticipatory planning, particularly when it comes to deceit. Deception is common in nature, but usually as a lifetime strategy, like the cryptic camouflage of the nightjar whose 'dead leaf' plumage enables it to rest safely on open woodland floor, or the Batesian mimicry of an edible butterfly which gains protection by its close resemblance to a noxious species. *Tactical deception* is rarer: this is where an act, which has one common, 'honest' meaning, is used in an unusual way that results in deceiving a conspecific (Byrne and Whiten 1985). As an illustration, consider a real observation of a young baboon that noticed an adult digging out a scarce and nutritious corm. The youngster, which had not been threatened, screamed; its mother ran into view, sized up the situation, and attacked the other adult. Both ran off – and the young baboon then ate the corm. Where this observation was made, underground foods like corms constitute a major source of winter nutrition, but young baboons are not strong enough to dig them out. The baboon's tactic is certainly based on deception, since the mother would only have behaved in the way she did if she misunderstood the true situation (i.e. mother baboons do not normally attack adults in order to gain foods for their offspring). We now know that all groups of monkeys and apes occasionally deceive in social circumstances; yet, in the wild, other species of animal have almost never been recorded doing so.

Is this illustration of primate tactical deception the result of creative thought? Or could the tactic have been learnt, by an individual who didn't understand the chain of logic so clear to any human observer? Suppose the young baboon had once before come across an adult, similarly excavating a nutritious corm, and approached it. Most likely, the adult would have threatened the youngster, who might well then have screamed in fear. However, when its mother came to the rescue, the corm would have been left available to the young animal. This reward would inevitably reinforce the act of screaming, making it likely to occur again in response to the same problem (roughly, 'adult with foodstuff that I want, and cannot get otherwise'). With sufficient lucky coincidences like this, the young baboon might also have learnt only to use the trick when the adult was lower rank than the mother, and when the mother was absent from view. In this particular case, associative learning of tactical deception is not implausible for a species whose members are able to learn from a single trial. Whiten and I, analysing a large corpus of records of deception in primates (Byrne 1997; Byrne and Whiten 1990; Whiten and Byrne 1988), found that we could construct this sort of 'plausible past history' to explain most cases as the result of learning – as long as the individuals could learn very quickly, as monkeys and apes can.

However, sometimes we found a case that was near impossible to explain this way. For instance, a dominant chimpanzee was observed confronted by an attempt at deception. Another chimpanzee inhibited its normal tendency to begin eating a coveted food item when it saw the dominant chimpanzee nearby (itself a tactic that could have been learned from past coincidences). The dominant's reaction showed that the deception was not successful: it hid and peeped out from behind a tree. Presumably thinking that the dominant had instead left, the subordinate chimpanzee picked up the food, and was then promptly relieved of it. Using the hide-and-peep tactic had effectively unmasked the other's deceit. The point is that hiding behind trees and peeping out is simply not in the normal range of chimp behaviour. This makes it difficult to see how that behaviour could possibly have been learned by associative conditioning: for a tactic to have been learned this way, it would have to have occurred previously, just when a deceit was being practised on the individual. It therefore seems more likely – more parsimonious, in the jargon of experimental psychology – that the animal already had a trace of suspicion, and used the hide-and-peep routine because it understood that the potential deceiver would give himself away if he thought he was unobserved. One single case would not be enough for such a strong claim, but in fact we found about 20 cases where 'insightless learning' became unparsimonious as an explanation, since such improbable coincidences had to be invented (Byrne and Whiten 1991). Interestingly, the cases of this type, where imaginary past histories became absurdly unlikely, all concerned great apes – chimpanzees, pygmy chimpanzees, gorillas and orangutans (Byrne and Whiten 1992). This was not because these species accounted for most deception in general – overall, we had more records of deception from monkeys than apes. Given this state of affairs, we concluded that great apes are capable of intentional deception.

The implication is that great apes, rather than having to rely on lucky series of coincidences to fashion their social trickery, are able to devise deceptive tactics by computation, calculation and planning: in other words, they are able to think.

INTERCOMMUNITY VIOLENCE

A much greater ability to anticipate the future is suggested by the unusual intercommunity interactions recorded in chimpanzees. In several different communities, over periods of years although not all the time, chimpanzees show behaviour that has been termed 'warfare' (Goodall *et al.* 1979; Nishida *et al.* 1985). Parties of male chimpanzees set off with apparent determination and patrol the peripheral areas of the community range. In contrast to the species' often noisy exuberance, they keep strikingly silent. Peripheral areas of one community's range overlap those of neighbouring communities, and when parties of strangers are encountered, the reaction depends on the balance of numbers. Large groups prompt noisy displays, given from a safe distance. However, when smaller groups are met, violence has been seen, directed largely at males although old females have also been attacked. The attacks are made with deliberation and persistence, often kept up for many minutes. The victims have been left severely wounded, and have generally died soon after. Younger females are coerced in the direction of the aggressors' core range, and have sometimes joined the party of males temporarily. In the longer term, in several cases these interactions have resulted in the destruction of a small community by a larger one, along with an expansion of the successful aggressors' range. As the males disappeared one by one, females transferred into the aggressors' community. The close parallel to Yanomami warfare has been noted (Boehm 1992; Chagnon 1974).

Genetically, these violent tactics do make sense: increasing the catchment area of resources and the number of females available to a group of successful aggressors. Male chimpanzees tend towards high relatedness, since their residence system is patrilocal: it is females that transfer between communities at adolescence (Nishida and Kawanaka 1972; Pusey 1979). Thus a high degree of mutual help among the males is not unexpected. But if the chimpanzee intercommunity violence is to be explained as kin selection, then it is difficult to avoid applying the same, genetical explanation to Yanomami actions: a 'gene for tribal warfare'. Alternatively, we know that humans can anticipate and plan for such consequences; perhaps chimpanzees can also do so. We cannot at present decide between these interpretations, but note that the chimpanzees' actions when on border patrol show that they can certainly anticipate consequences in the shorter-term future. This is shown when female chimpanzees with babies, or human researchers, are occasionally allowed to accompany the border patrols (Goodall 1986; Goodall *et al.* 1979). Any noise they make is immediately greeted with threats by the males: chimpanzee mothers rapidly comfort noisy babies, people take care not to break twigs. Yet there is no plausible history of past experience that might have conditioned the males to associate noise with danger. Border patrols are relatively infrequent events so there can be little opportunity for shaping behaviour by reinforcement, and when – as here – the 'punishment' for error is death, no trial and error learning can result. It would seem more likely that the males anticipate the effect that noise would have on groups from the neighbouring community, using their understanding of the situation: in other words, thinking it out.

PREDATOR CONTROL

More speculatively, a single incident of chimpanzee behaviour towards leopards raises the possibility of even greater foresight. In this episode, a group of old male chimpanzees cornered a mother leopard in her breeding den, a long and narrow cave (Byrne and Byrne 1988; Hiraiwa-Hasegawa *et al.* 1986). After many violent displays – and retreats whenever she roared – one individual went entirely into the cave, and emerged with the cub, which was estimated to be only 2–3 months old. The males bit and kneaded it until it was dying, then abandoned it, without any attempt at eating the meat.

Killing a cub is precisely what humans would do to discourage a mother leopard from remaining in the area, and also to attempt to reduce the overall population of leopards. People can 'think through' the likely results of such an action, and decide if the probable gains are worth the certain risks. Chimpanzees are known to suffer predation by leopards, and their hair has been recorded in leopard scats. However, leopards mainly hunt at night, and in daylight chimpanzees can usually drive away leopards. To judge by accounts of man-eating and gorilla-eating leopards (Corbett 1948; Schaller 1963), the greatest risk is of a single individual specialising in great ape predation. Nevertheless, to enter a narrow cave occupied by a big cat and snatch its infant would appear so crazily dangerous that the chimpanzees must have had a strong motivation to carry it out, an interpretation supported by the persistence of their attack. Meat was not eaten, and in the absence of any immediate reward, their behaviour cannot be explained as conventional animal learning. Dismissing it as an aberrant, pathological act does not make sense: although the majority of the large chimpanzee group appeared as scared as the researchers observing them, at the thought of a cornered mother leopard desperate in defence of her young, the old males pursued their attack relentlessly for over an hour before they met with success. As always, any behavioural trait can potentially evolve, if it increases the relative genetic representation of those who possess it among the population. However, a genetical explanation is usually suspected when a behaviour is universal across different populations, and to evoke a chimpanzee gene for leopard-baby killing to explain a single observation is very weak. Perhaps the 'least implausible' explanation is that the chimpanzees had an understanding of the likely effects of their actions in the future.

EXPERIMENTAL DIAGNOSIS OF THOUGHT?

It would be satisfactory to be able to investigate such a difficult question as animal thinking experimentally, rather than have to rely on *post hoc* interpretation of observational data. Unfortunately, no experimental test of thinking exists. The closest that experimentalists have so far approached is to set up problems that would certainly benefit from thought, in the hope that the actual pattern of solution 'looks as if' thinking were involved.

This approach was famously taken by Wolfgang Köhler, in studies undertaken while he was on Tenerife (Köhler 1925). He gave chimpanzees various novel problems, including that of raking in a banana, when all the sticks provided were too short. In one observation, a chimpanzee, Sultan, gave up trying this puzzle after several attempts; ten minutes later, the keeper excitedly came to Köhler, recounting how Sultan had suddenly got up from where he had been playing with the sticks, confidently inserted one stick in another, hollow stick, then had gone straight over to the banana and raked it in.

Köhler's observation is often cited as evidence of 'insight': although the task has been abandoned, thinking goes on, and a sudden imaginative leap brings the problem to the ape's mind, solved. People resonate to this story, and readily report their own cases, when quite different behaviour is interrupted by a sudden 'Ah ha!' But notice the behaviour that was 'interrupted': playing with sticks. This is not an irrelevant activity.

In using think-aloud protocols to study human problem solving, I found that the excitement of noticing a solution to an abandoned problem caused people to forget what they'd been thinking about just beforehand (Byrne 1977). On the recorded protocol, it was always clear that their just-prior thoughts were closely related to their 'flashes of insight'. But the subjects themselves imagined the idea had 'just suddenly come to them', as if it were a result of unconscious thinking. I suggest that in both chimpanzee and human cases, appropriate solutions are not necessarily computed by a continuing process of directed thought, conscious or otherwise, but may be *noticed* when they occur in current activity. Kekulé is supposed to have dreamed about snakes grasping each other's tails, and so discovered the structure of benzene. If great scientists' insights are derived in ways like this, then the same applies: noticing, not insight, may be the mechanism, and the triviality or greatness of the process is only a matter of the degree of abstraction of the leap. That is, the hierarchical level of description differs at which the two things are 'the same': low-level abstraction for stick-in-play to stick-for-task, high-level abstraction for snakes to molecules. Unfortunately, we do not understand what goes on in 'noticing' even in humans, so this is little help for interpreting animal actions.

CONCLUSIONS

On the basis of the disparate strands of evidence reviewed here, I would argue that the crucial ability underlying the various actions of great apes – but not monkeys or most other species of animal – is best identified with human non-verbal thinking. This process is computational, and involves some sort of reasoning with representations of reality, of states desired in the future, or of beliefs of other individuals. From the distribution of the evidence across modern species, the ability to think was most likely present in the common ancestor of all great apes, around 16 mya, but not before (Byrne 1995). Great ape thinking appears adaptive, in that it enables them to anticipate and prepare for problems at least in the immediate future and perhaps over longer time scales. On the basis of current evidence, their thinking consists of making novel connections among known pieces of knowledge, and hence deriving novel solutions. As in humans, apparently 'insightful' solutions may result from *noticing* connections to persisting problems in the current stream of experience, rather than long sequences of deliberate planning.

At present, this argument is not completely compelling. No single source of evidence forces us to accept claims that any non-human can think in a propositional and representational way. No experimental test of thinking in animals has yet been devised. However, there is a *convergence* of evidence towards a clear cognitive difference between monkeys and apes. Both groups of animals are highly social, confronted by broadly similar environmental problems (often in the same sites), and often achieve similar ends by their behaviour. But despite abundant evidence of rapid learning in monkeys, there is no sign of more than this. In great apes, researchers are continually confronted by small signs that the animals' behaviour is significantly more like our own: the first glimmerings of creative thinking.

NOTES

1 Sometimes the term 'thought' is used synonymously with 'awareness' (Griffin 1984). However, in this chapter I am taking instead the sense that is closer to 'planning' or 'computation', as would be normal in cognitive psychology.

2 In addition, it has recently been discovered that one population of orang-utans also makes and uses tools, of two types, but these habits have yet to be studied as intensively as the chimpanzee skills (van Schaik *et al.* 1996).

REFERENCES

Boehm, C. (1992) 'Segmentary "warfare" and the management of conflict: comparison of East African chimpanzees and patrilineal-patrilocal humans.' In A.H. Harcourt and F.B.M. de Waal (eds) *Coalitions and Alliances in Humans and Other Animals*, pp. 137–173. Oxford: Oxford University Press.

Boesch, C. and Boesch, H. (1981) 'Sex differences in the use of natural hammers by wild chimpanzees: a preliminary report.' *Journal of Human Evolution* 10: 585–593.

Boesch, C. and Boesch, H. (1983) 'Optimisation of nut-cracking with natural hammers by wild chimpanzees.' *Behaviour* 26: 265–288.

Boesch, C. and Boesch, H. (1984) 'Mental map in wild chimpanzees: an analysis of hammer transports for nut cracking.' *Primates* 25: 160–170.

Byrne, R.W. (1977) 'Planning meals: problem-solving on a real data-base.' *Cognition* 5: 287–332.

Byrne, R.W. (1995) *The Thinking Ape: Evolutionary Origins of Intelligence*. Oxford: Oxford University Press.

Byrne, R.W. (1997) 'What's the use of anecdotes? Attempts to distinguish psychological mechanisms in primate tactical deception.' In R.W. Mitchell, N.S. Thompson and L. Miles (eds) *Anthropomorphism, Anecdotes, and Animals: The Emperor's New Clothes?*, pp. 134–150. New York: SUNY Press Biology and Philosophy.

Byrne, R.W. and Byrne, J.M. (1988) 'Leopard killers of Mahale.' *Natural History* 97: 22–26.

Byrne, R.W. and Whiten, A. (1985) 'Tactical deception of familiar individuals in baboons (*Papio ursinus*).' *Animal Behaviour* 33: 669–673.

Byrne, R.W. and Whiten, A. (1990) 'Tactical deception in primates: the 1990 data-base.' *Primate Report* 27: 1–101.

Byrne, R.W. and Whiten, A. (1991) 'Computation and mindreading in primate tactical deception.' In A. Whiten (ed.) *Natural Theories of Mind*, pp. 127–141. Oxford: Basil Blackwell.

Byrne, R.W. and Whiten, A. (1992) 'Cognitive evolution in primates: evidence from tactical deception.' *Man* 27: 609–627.

Chagnon, N.A. (1974) *Studying the Yanomamo*. New York: Holt, Rinehart & Winston.

Corbett, J. (1948) *The Man-Eating Leopard of Rudraprayang*. Oxford: Oxford University Press.

Dawkins, M.S. (1993) *Through our Eyes Only? The Search for Animal Consciousness*. Oxford: W.H. Freeman.

Dennett, D.C. (1983) 'Intentional systems in cognitive ethology: the "Panglossian paradigm" defended.' *Behavioural and Brain Sciences* 6: 343–390.

Dickinson, A. (1980) *Contemporary Animal Learning Theory*. Cambridge: Cambridge University Press.

Galef, B.G. (1988) 'Imitation in animals: history, definition, and interpretation of data from the psychological laboratory.' In T. Zentall and B.G. Galef, Jnr (eds) *Social Learning: Psychological and Biological Perspectives*, pp. 3–28. Hillsdale, NJ: Erlbaum.

Goodall, J. (1986) *The Chimpanzees of Gombe: Patterns of Behavior*. Cambridge, MA: Harvard University Press.

Goodall, J., Bandora, A., Bergmann, E., Busse, C., Matama, H., Mpongo, E., Pierce, A. and Riss, D. (1979) 'Intercommunity interactions in the chimpanzee population of the Gombe National Park.' In D. Hamburg and E.R. McCown (ed.) *The Great Apes*, pp. 13–54. Menlo Park, CA: Benjamin Cummings.

Goodall, J. v. L. (1973) 'Cultural elements in a chimpanzee community.' In E.W. Menzel (ed.) *Precultural Primate Behaviour*, pp. 144–184. Basel: Karger.

Green, S. (1975) 'Dialects in Japanese monkeys: vocal learning and cultural transmission of locale-specific vocal behaviour?' *Zeitschrift fur Tierpsychologie* 38: 304–314.

Griffin, D.R. (1976) *The Question of Animal Awareness: Evolutionary Continuity of Mental Experience*. New York: Rockefeller University Press.

Griffin, D.R. (1984) *Animal Thinking*. Cambridge, MA: Harvard University Press.

Hiraiwa-Hasegawa, M., Byrne, R.W., Takasaki, H. and Byrne, J.M. (1986) 'Aggression towards large carnivores by wild chimpanzees of Mahale mountains National Park, Tanzania.' *Folia Primatologica* 47: 8–13.

Köhler, W. (1925) *The Mentality of Apes*. London: Routledge & Kegan Paul.

Miller, G.A., Galanter, E. and Pribram, K. (1960) *Plans and the Structure of Behavior*. New York: Holt, Rinehart & Winston.

Newell, A., Shaw, J.C. and Simon, H.A. (1958) 'Elements of a theory of human problem solving.' *Psychological Review* 65: 151–166.

Newell, A. and Simon, H.A. (1972) *Human Problem Solving*. New York: Prentice-Hall.

Nishida, T. (1986) 'Local traditions and cultural transmission.' In B.B. Smuts, D.L. Cheney, R.M. Seyfarth, R.W. Wrangham and T.T. Struhsaker (eds) *Primate Societies*, pp. 462–474. Illinois and London: University of Chicago Press.

Nishida, T. and Kawanaka, K. (1972.) 'Inter-unit-group relationships among wild chimpanzees of the Mahali Mountains.' *Kyoto University African Studies* 7: 131–169.

Nishida, T., Hiraiwa-Hasegawa, M., Hasegawa, T. and Takahata, Y. (1985) 'Group extinction and female transfer in wild chimpanzees in the Mahale National Park, Tanzania.' *Zeitschrift fur Tierpsychologie* 67: 284–301.

Pusey, A. (1979) 'Intercommunity transfer of chimpanzees in Gombe National Park.' In D.A. Hamburg and E.R. McCown (ed.) *The Great Apes*, pp. 465–479. Menlo Park, CA: Benjamin Cummings.

Ristau, C. (1991) 'Aspects of the cognitive ethology of an injury-feigning bird, the piping plover.' In C. Ristau (ed.) *Cognitive Ethology: The Minds of Other Animals*, pp. 91–126. Hillsdale, NJ: Erlbaum.

Roper, T.J. (1983) 'Learning as a biological phenomenon.' In T.R. Halliday and P.J.B. Slater (eds) *Animal Behaviour*, vol. 3, pp. 178–212. Oxford: Blackwell Scientific Publications.

Schaller, G.B. (1963) *The Mountain Gorilla*. Chicago: Chicago University Press.

Spence, K.W. (1937) 'Experimental studies of learning and higher mental processes in infra-human primates.' *Psychological Bulletin* 34: 806–850.

Terkel, J. (1994) 'Social transmission of pine cone feeding behaviour in the black rat.' In B.G.J. Galef, M. Mainardi and P. Valsecchi (eds)*Behavioural Aspects of Feeding*, pp. 229–256. New York: Harwood Academic.

Tomasello, M. (1990) 'Cultural transmission in the tool use and communicatory signaling of chimpanzees?' In S.T. Parker and K.R. Gibson (eds) *'Language' and Intelligence in Monkeys and Apes*, pp. 274–311. Cambridge: Cambridge University Press.

van Schaik, C.P., Fox, E.A. and Sitompul, A.F. (1996) 'Manufacture and use of tools in wild Sumatran orangutans. Implications for human evolution.' *Naturwissenschaften* 83: 186–188.

Visalberghi, E. and Fragaszy, D. (1990). 'Food-washing behaviour in tufted capuchin monkeys (*Cebus apella*) and crab-eating macaques (*Macaca fascicularis*).' *Animal Behaviour* 40: 829–836.

Visalberghi, E. and Limongelli, L. (1994) 'Lack of comprehension of cause–effect relationships in tool-using capuchin monkeys (*Cebus apella*).' *Journal of Comparative Psychology* 103: 15–20.

Visalberghi, E. and Limongelli, L. (1996) 'Acting and understanding: tool use revisited through the minds of capuchin monkeys.' In A.E. Russon, K.A. Bard and S.T. Parker (eds) *Reaching into Thought: The Minds of the Great Apes*, pp. 57–79. Cambridge: Cambridge University Press.

Whiten, A. and Byrne, R.W. (1988) 'Tactical deception in primates.' *Behavioural and Brain Sciences* 11: 233–273.

CHAPTER EIGHT

'HOMO': THE CREATIVE GENUS?

MARK LAKE

Not *H. 'faber'* or *H. 'sociologicus'*, but *H. 'creatrix'*? Creativity is perhaps the single phenomenon that most clearly distinguishes humans from other primates; but at what point in human evolution did creativity appear? *A priori* there appear to be two possibilities. Perhaps ours is a creative genus, in which case the first signs of creativity should be sought with the appearance of early *Homo* around 2 mya. Alternatively, creativity might be the defining property of modern humans – our own subspecies *H. sapiens sapiens* – who evolved within the last 200,000 years. This chapter addresses the first possibility: it considers the evidence for creative thought in early *Homo*.

EARLY *HOMO*

On the basis of genetic evidence, it is thought that the hominid and ape lineages split at the very end of the Miocene, about 6 mya (Sibley and Ahlquist 1984). The earliest fossil hominids date between 4.5–3.9 mya. Found in east Africa, *Australopithecus* (*Ardipithecus*) *ramidus* and *Australopithecus anamensis* were apelike and lived in wooded or bush environments (White *et al.* 1994; WoldeGabriel *et al.* 1994). By 4 mya, fossils of *Australopithecus afarensis* provide evidence for distinctively hominid characteristics such as bipedalism and reduced dentition (Johanson and Edey 1980). The gradual loss of apelike features continued with the appearance of *Australopithecus africanus* one million years later, and accelerated with the appearance of *Homo* at *c.*2 mya (Chamberlain and Wood 1987; Johanson and White 1979). It should not be supposed, however, that early hominid evolution led exclusively to the arrival of our genus. Instead, the first species of *Homo* were members of an adaptive radiation that also included at least two 'robust australopithecine' hominid species, *Paranthropus boisei* and *Paranthropus robustus* (Wood 1992a), characterised by a heavily built physique, highly developed facial muscles and very large molar teeth.

The taxonomy and phylogeny of early *Homo* is currently the subject of considerable debate (e.g. Chamberlain and Wood 1987; Groves 1989; Johanson *et al.* 1987; Skelton *et al.* 1986; Wood 1991), but there is a general consensus that the first members of our genus represent an important step in human evolution. Early *Homo* is characterised morphologically by an enlarged braincase, which marks the beginning of a trend to rapidly increasing encephalisation in later species ancestral to *H. sapiens* (Figure 8.1; Aiello 1996). The expansion of the cranial vault to a capacity typically in excess of 700 ml results in a notably more vertical face. In addition, the lower jaw is relatively light and may even include a slight chin; dentition is similarly reduced. Although there is little post-cranial evidence for early *Homo*, it is thought to have been essentially bipedal (Bilsborough 1992). In addition to these significant morphological changes, the archaeological record provides evidence for a behavioural shift towards greater complexity in subsistence strategy and tool-use. In particular, it is early *Homo* who is conventionally associated with Oldowan stone tools (ibid.). Similarly, it is early *Homo* who is associated with the incorporation of meat into the diet and perhaps also with more differentiated patterns of land-use exhibiting a high degree of spatial redundancy (Foley 1987).

The earliest members of the genus Homo are variously attributed to one, two or three species: *H. habilis*, *H. rudolfensis* and *H. ergaster*. *H. habilis* and *H. rudolfensis* both appear by 1.9 mya and are differentiated by several putative markers of interspecific variability (Wood 1991). *H. habilis* is geographically more widespread (it includes fossil OH 13 from Olduvai Gorge and fossil KNM-ER 1813 from Koobi Fora, both in east Africa) and is characterised by a smaller braincase, small protruding face and small jaw with more human-like dentition. In contrast, *H. rudolfensis* may be geographically more restricted (the species attribution is reserved for the larger fossils from Koobi Fora, such as KNM-ER 1470) and is characterised by a large braincase with a broad flat face, a large jaw and large teeth. *H. ergaster* appears by 1.78 mya and represents the earliest east African fossils generally regarded as belonging to the *H. erectus* grade (ibid.). It is characterised by a brain size of 750 ml or more, a wide cranial base, a relatively vertical cranial vault and essentially human-like dentition.

For the purposes of the present enquiry, no distinction is made between *H. habilis*, *H. rudolfensis* and *H. ergaster*. In one sense this is problematic because there is always a possibility that each species possessed a unique set of cognitive abilities, something that is difficult to assess given that we do not know which of them were responsible for the archaeological record. Indeed, the very fact that *H. habilis* and *H. rudolfensis* may have been genuinely sympatric (Wood 1992b) supports the notion that their behaviours and underlying cognitive abilities differed sufficiently to prevent niche overlap. Nevertheless, it is possible to justify treating all three species of early *Homo* together because from a wider perspective they do appear to represent a common trend towards a more human

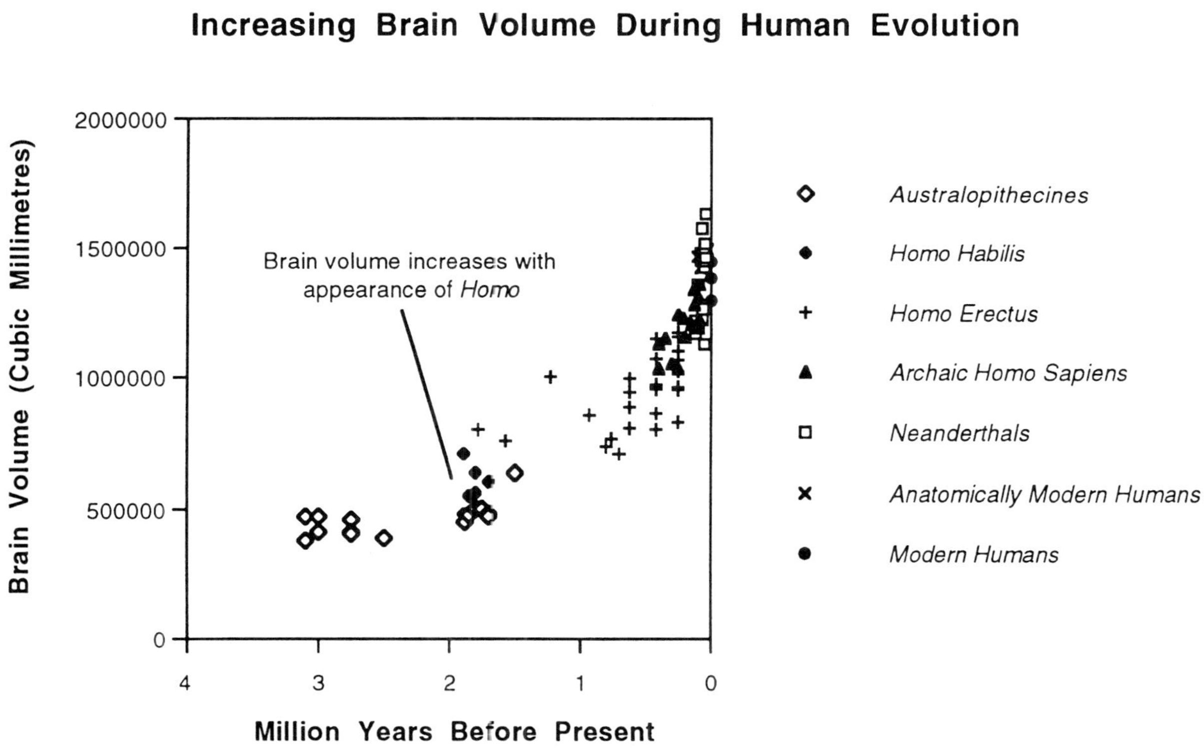

Figure 8.1 Increasing brain volume during human evolution
Source: Aiello and Dunbar 1993

adaptation. This is especially clear when their relatively gracile morphologies are contrasted with the sexual dimorphism and megadonty of the robust australopithecines. And, perhaps more importantly, their increased brain size and reduced anterior dentition suggest that all three species adopted an increased reliance on high-quality foodstuffs to overcome dry season scarcity, in direct contrast to the alternative strategy of bulk intake that appears to have been adopted by the robust australopithecines (Foley 1987; Grine 1981; Shipman and Harris 1988).

INFERRING CREATIVITY

What would convince us that early *Homo* was creative? Any attempt to answer this question requires a definition of creativity. Margaret Boden (1994) recognises two types of creativity and suggests that instances of each may be further subdivided according to whether or not they are original in an absolute sense. The principal distinction she proposes is between *improbabilist* creativity and *impossibilist* creativity. Improbabilist creativity involves making new connections between facts and ideas, but these connections are made within existing conceptual spaces. We are impressed by instances of improbabilist creativity because in retrospect it appears unlikely that anyone would make the new connections. Impossibilist creativity is perhaps more profound. It involves the transformation of existing conceptual spaces to permit new thoughts that were previously unthinkable.

If instances of creativity differ in whether they permit new ideas that were unlikely or new ideas that were impossible, they also differ in whether or not they are original in an absolute sense. For this reason, Boden further distinguishes between *psychological*, or P-creativity, and *historical*, or H-creativity. All creative thoughts are instances of P-creativity. If you connect ideas or facts that you did not previously think were connected, or if you transform your conceptual space then you have been psychologically creative. Nevertheless, your creative thoughts will arouse wider interest if they are original in the sense of never having been thought before by anyone else. Only in this latter case will you have exercised historical creativity.

If one accepts Boden's definitions of creativity then evidence that members of early *Homo* made novel connections between ideas and facts, or thought the previously unthinkable, would count as evidence that ours is indeed a creative genus. This is adequate in principle, but in practice cognitive connections and conceptual spaces are simply not the sort of things that we can dig up. It follows that we are unlikely ever to know for certain whether early *Homo* was creative *sensu* Boden. Nevertheless, it does not follow that we should avoid the question altogether. Merely asking whether early *Homo* was creative forces us to consider what hominids were doing mentally to produce the behaviours that created the archaeological evidence. This is, in fact, a similar strategy to that pursued by cognitive ethologists. For example, there is no way of being absolutely certain what a Machiavellian chimpanzee is thinking as it conceals sexual arousal from a higher ranking competitor (de Waal 1982). Nevertheless, by careful observation, cognitive ethologists can demonstrate that the most plausible cognitive explanation for this behaviour is an intention to make the competitor believe something that is not true (Whiten and Byrne 1988). Only the most diehard Skinnerian would argue that cognitive ethology has nothing to contribute to our understanding of animal behaviour. (For a critique of Skinner's programme, see Chomsky 1959.) Consequently, it is worth considering just how far one can pursue the analogy between inferring the thoughts of long-dead hominids and the practice of cognitive ethology.

Perhaps the most striking difference between the archaeological and ethological inference of thought is the additional degree of control available to the latter. There are numerous experiments in cognitive ethology that demand repetition with altered conditions, and there are likewise many anecdotal observations that permit inference only in the context of longitudinal studies. For example, if a culturally learned behaviour is one that, among other criteria, was not already present in the repertoire of the learner (Byrne 1995), then cultural learning can be demonstrated only if the learner has been the subject of continuous observation (Whiten and Ham 1992). Whereas cognitive ethologists can, at least in principle, continuously observe the behaviour of their subjects and, if necessary, alter the context in which they operate, these strategies are simply not available to archaeologists. It is possible to view the past as a laboratory in which behaviour can be studied in the long term against a backdrop of different social and natural environments, but the evidence almost invariably pertains to groups rather than specific individuals.

Boden's view that the source of creativity ultimately resides in individuals – even when it has historical significance – has mixed implications for archaeological inference. On the one hand, it might appear that archaeologists cannot document psychological creativity because they rarely, if ever, have access to longitudinal data referable to *specific* individuals. In the absence of time transgressive data, how can one determine whether the acquisition of a novel behaviour resulted from teaching or from individual creativity? On the other hand, historical creativity is, at least in principle, amenable to observation: a novel behaviour generated by creative thought must have a first appearance in the archaeological record (whether the observed instance was genuinely the first occurrence is to some extent trivial). Since historical creativity is nothing other than a special case of psychological creativity, it follows that the latter ought to be archaeologically demonstrable, at least *per se*. What is less amenable to archaeological investigation is the extent to which individuals were routinely creative in the purely psychological sense.

These inferential considerations suggest a further dimension to creativity that is not adequately conceptualised in Boden's framework, at least for archaeological purposes. Her computational approach dictates that she is primarily concerned with creativity as a cognitive phenomenon; however, the extent to which such creativity becomes widely manifest and thus observable depends on the channels of communication in which it is situated. This latter might be an irrelevance to cognitive ethologists who are in a position to strip away the effects of sociality, but it is unavoidable for archaeologists who largely deal with population-level phenomena. In other words, we must recognise that asking whether early *Homo* was creative implies two questions, not one: was early *Homo* cognitively capable of creative thought, and did early *Homo* live in a 'culture' of creativity? These questions are addressed in the context of the two spheres of activity for which we have the most archaeological evidence: resource acquisition and stone tool manufacture.

MOVING BONES AND STONES: CREATIVITY IN RESOURCE ACQUISITION?

The archaeological evidence for the behaviour of early *Homo* essentially comprises scatters of flaked stone tools and fossilised animal bones eroding out of sediments at a number of locations along the African Rift Valley (Gamble 1993). To date, evidence has been found as far north as Hadar in Ethiopia and as far south as Peninj in Tanzania. Although the following discussion is based specifically on the evidence from Koobi Fora in northern Kenya, it can be generalised to other locations. The stone tools and faunal remains at Koobi Fora are scattered across the landscape in thin ribbons where erosion has cut through the sediment in which they were originally deposited (Isaac *et al.* 1981). The density of artefacts ranges from less than one per 25 m^2 – the 'scatters' – through intermediate patches of between 2–20 artefacts per 25 m^2 in a restricted area of up to 500 m in cross section – the 'mini sites' – to peak levels of over 100 artefacts per 25 m^2, sometimes clustering so that there are over 1,000 in an area with a diameter of 10–30 m – the 'sites' (ibid.). The sites and mini sites have been classified into four types on the basis of their contents (Isaac 1984). Type A sites are concentrations of stone tools without bone, or where bone is present only at normal background densities. Type B sites are clusters of stone tools intermingled with bones representing substantial parts of a single animal. Type C sites are concentrations of stone tools that are interspersed among bones deriving from numbers of individual animals belonging to different species. Finally, type M sites are concentrations of bones bearing cutmarks, but lacking stone tools.

During the 1960s and 1970s, researchers interpreted the variability in the density and composition of archaeological material as evidence that early *Homo* pursued resource acquisition strategies as complex as those of modern hunter-gatherers. In particular, it was supposed that sites of type A, B and C represented 'workshops', 'butchery sites' and central places respectively (e.g. Clark 1970; Isaac 1971; Leakey 1971). This view was formalised in Glynn Isaac's Home Base Hypothesis. Building on the notion that type C sites provided evidence for use of a central place, Isaac proposed that the behaviour of early *Homo* included food sharing and a division of labour (Isaac 1978). He suggested that animal foods, obtained by males, and plant foods, obtained by females, were redistributed at a home base, which provided a safe haven for rearing young and was also the site of stone tool manufacture. In the early 1980s, the Home Base Hypothesis was strongly challenged by Lewis Binford. He argued that Isaac had paid insufficient attention to the role of non-hominid agency in site formation (Binford 1981). Once the activities of carnivores were taken into account, along with the effects of natural processes such as water flow, it seemed to him more plausible that early *Homo* was simply a marginal scavenger who moved around the landscape in a manner similar to the routed foraging practised by modern apes. Today, however, the consensus is that neither Binford nor Isaac was entirely correct in their reconstructions of resource acquisition by early *Homo* (Figure 8.2).

More than a decade of taphonomic and actualistic research suggests that early *Homo* was neither a marginal scavenger nor exclusively a hunter. Instead, the sequence and location of carnivore tooth marks and stone tool cutmarks on bones point to a strategy of opportunistic meat acquisition (Potts 1988). Early *Homo* may have hunted small or juvenile herbivore species, but was probably restricted to scavenging the carcasses of larger species. This interpretation militates against the Home Base Hypothesis because it suggests that meat was not routinely acquired in large enough quantities to permit regular sharing. A further objection to the Home Base Hypothesis is that a site littered with discarded bones would attract carnivores and thus not be safe for rearing young or for sleeping (Binford 1983). Richard Potts has attempted to reconcile the apparently opportunistic nature of meat acquisition by early *Homo* with the fact that certain places in the landscape do seem to have been visited repeatedly. His Stone Cache Hypothesis (Potts 1984, 1988) was founded on a consideration of the energetic costs of carrying and using stone tools and unworked raw materials in the process of foraging. A simple computer simulation suggested that the least costly way of bringing stones and carcasses together for processing would be to redistribute both resources to a common locality. Potts envisaged the transport of carcasses to the nearest cache of stones. He stressed that the origin of such caches might simply be stones dropped at locations where their transport and use was incidental, but over time, as debris accumulated around these points, they became remembered focal points for transported stones.

Although Potts' Stone Cache Hypothesis is not without its problems (see Mithen 1991), it serves as a useful summary of the most significant

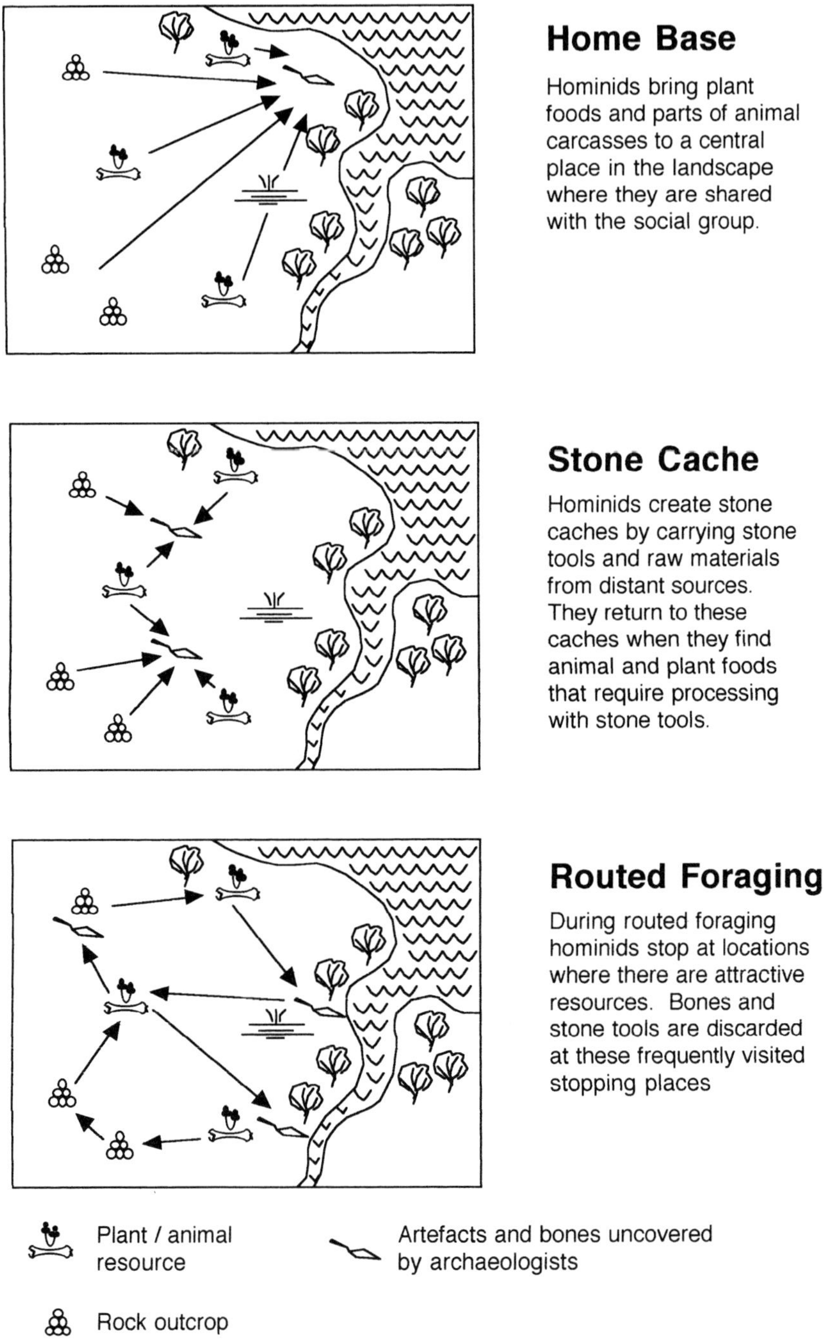

Figure 8.2 Possible hominid resource acquisition strategies
Source: Adapted from Jones *et al.* 1992

aspects of the resource acquisition strategies pursued by early *Homo*. Thus the question that concerns us here is whether the opportunistic acquisition of meat and the repeated use of particular locations constitutes evidence for creative thought. It is worth considering each aspect of resource acquisition in turn.

The opportunistic acquisition of meat does not necessarily imply creative thought, merely flexibility. If the opportunistic acquisition of meat is the conceptual space, then this would not be transformed in any given episode of scavenging or hunting. Likewise, there would be no *necessary* requirement for new connections between ideas and facts, except in the most trivial sense of accommodating new facts (for example the discovery of a carcass), which hardly constitutes a uniquely human ability. Nevertheless, while the opportunistic acquisition of meat might not be creative in the impossibilist sense, it might have created a new context for – although not required – improbabilist creativity. One can, for example, envisage a scenario in which a member of early *Homo* made the novel connection between circling vultures and the presence of a carcass. What would be creative about this in Boden's improbabilist sense is not the linking of the vultures and the carcass in the specific instance, but the creation of a new rule for prediction. Such a rule could have been generated once, or repeatedly by different hominids. Unfortunately, there is no direct evidence for or against the use of specific natural cues by early *Homo*, let alone for the frequency of psychological creativity in recognising them.

The repeated use of particular locations outwardly appears rather lacking in creativity, but this is a somewhat superficial reading. If one assumes that the pattern of land-use by early *Homo* was not the subject of a genetic predisposition, then a pattern involving spatial redundancy represents a choice on the part of an individual or a group of individuals. The question that naturally follows is whether differing patterns of organisation such as routed foraging and stone caching can be accommodated within one conceptual space, or whether the change from one to the other requires a transformation of the pre-existing conceptual space. In other words, does it require impossibilist creativity? It is not clear on what basis this question could be answered. According to Boden (1994), individuals can transform their conceptual spaces only if they possess reflexive descriptions of their own procedures and ways of varying them; in the absence of such descriptions, they are limited to exploration of their conceptual spaces. It follows that the adoption of a new pattern of land-use could have resulted from a transformation of conceptual space only if early *Homo* possessed reflexive descriptions of the generative rules by which it organised land-use. If early *Homo* did not possess such reflexive descriptions then, by definition, the switch from one pattern of land-use to another must have been achieved simply through application of the existing generative rules. Unfortunately, the evidence for land-use

(Potts 1994) does not allow us to deduce whether early *Homo* possessed reflexive descriptions of the rules used to generate the pattern of land-use. Ultimately, there is no way of determining whether variability in land-use is referable to impossibilist creativity because we have no means to identify the limits of the relevant conceptual spaces.

REMOVING FLAKES: CREATIVITY IN STONE TOOL MANUFACTURE?

The first stone tools date between 3 and 2 mya and are referred to as the Omo industrial complex (Hall *et al.* 1985; Harris and Johanson 1983). They occur in the Shungura Formation in the Omo area of Ethiopia and also at Lokalalei, where they have been found in the Nachukui Formation of West Turkana, Kenya. The tools from Omo are essentially smashed pieces of quartz and are often difficult to distinguish from naturally broken rock. It is possible that technical skill is masked by the fracture dynamics of the raw material. The artefacts from Lokalalei, however, were manufactured from much more easily worked raw material (Kibunjia 1994). Despite this, they are very rudimentary, even when compared with the subsequent Oldowan industry (Figure 8.3). Furthermore, the excavator's analysis showed that about 80 per cent of attempts to produce flakes had failed, leaving characteristic step fractures on the cores.

By 1.8 mya, hominids were producing a wider range of tool forms than is evidenced in the Omo industries. These new tools make up the Oldowan industrial complex (Leakey 1971) and are found until *c.*1.5 mya. Oldowan artefacts were mostly made from volcanic lava, notably basalt and quartz, although other materials such as chert and gneiss were occasionally used (Potts 1988). They are classified into four broad categories (Schick and Toth 1993). Heavy-duty tools are cores made of cobbles or chunks of rock. Some are named after their supposed functions, such as choppers and heavy-duty scrapers, others such as polyhedrons are named after their overall shape. Light-duty tools are made from flakes and smaller chunks of rock, and are characterised by the presence of retouch – the removal of small flakes to provide a useful edge for some purpose. Light-duty tools are named after their supposed function, for example light-duty scrapers, awls and burins. Utilised pieces are artefacts that have not been deliberately flaked into a specific form but have been shaped as a result of their use for some task. They include flakes with chipped edges, and hammerstones and anvils. The final class of Oldowan artefacts is debitage: the fine flakes and fragments that represent the 'waste' from tool manufacture.

Given that more than one species of hominid lived in the period 1.8–1.5 mya, we cannot be certain who actually manufactured the Oldowan tools. It has been argued that the robust australopithecines were tool makers, on the grounds that they possessed the requisite manual dexterity (Susman 1991), and recent evidence pushing the date for the

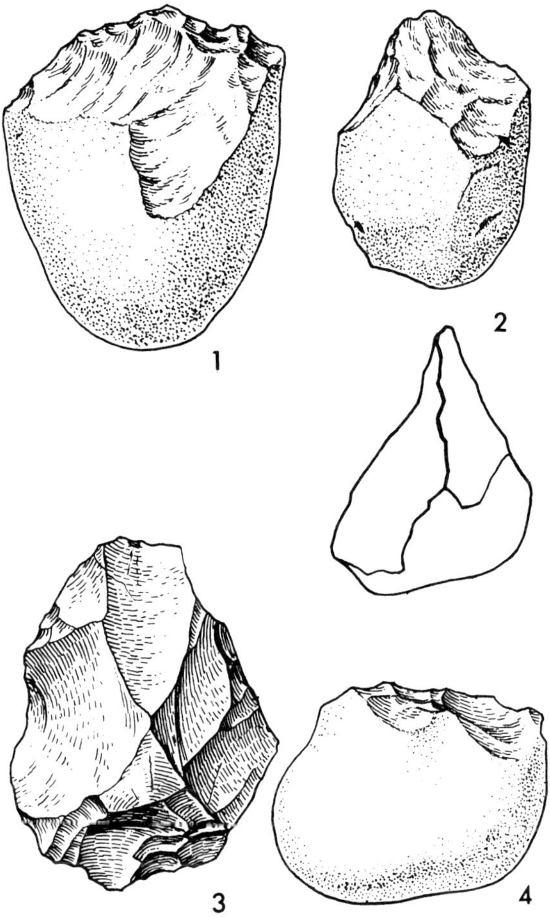

Figure 8.3 Typical Oldowan artefacts. (1) Chopping tool from Olduvai Bed I, (2) Proto-Biface from Olduvai Bed II, (3) Tool from Sterkfontein, (4) Chopping tool from Vallonet Cave.

Oldowan back to 2.6–2.5 mya (Semaw *et al.* 1997) may provide further support for this notion (Wood 1997). Usually, however, it has been assumed that *Homo* was the tool maker, largely on the rather circular logic that increased brain size should correlate with more complex tool-use. Here it is assumed that early *Homo* did make Oldowan tools; whether other hominids did likewise is less relevant for present purposes. Tool manufacture by early *Homo* might have been creative in at least two ways. It could be that the use of stone for tool manufacture was in itself creative. Alternatively, perhaps the production of new tool forms was creative.

Hominids are unique in the use of stone for tool manufacture. Other animals *use* stone tools, but none spontaneously *make* stone tools. For example, mud wasps and California sea otters use unmodified stones as tamps and anvils (Schick and Toth 1993). Wild chimpanzees also use unmodified stones as both hammers and anvils (Boesch and Boesch 1983). Only chimpanzees actually make stone tools, and even then only as a direct result of encouragement in captivity (Toth *et al.* 1993). Is there any sense, then, in which early *Homo*'s use of stone for tool manufacture could be deemed creative? Following Boden's scheme, does the use of stone imply the transformation of existing conceptual spaces, or perhaps alternatively the construction of new connections between facts about stone and the idea of a made tool? There is no straightforward archaeological route to answering this question, but some insight can be gained from consideration of chimpanzee tool manufacture and use.

Wild chimpanzees are routine tool makers (McGrew 1992): for example, the deliberate and careful manufacture of termite fishing wands is well documented (ibid.). Given the equally well documented use of stones, it is perhaps revealing that they do not manufacture stone tools. This discrepancy suggests that stone tool manufacture does require the transformation of conceptual space. Although poorly documented, it is not unreasonable to assume that the use of stone hammers must occasionally result in the production of chunks or flakes. Such an event would provide new information about stone: that it is a potential source of sharp edges. What is striking is that wild chimpanzees *never* connect this new information with their existing ideas about tools. It seems that, in the wild, stone tool manufacture is not only improbable, but impossible. This contrasts with the evidence from captivity that chimpanzees can manufacture rudimentary stone tools if given sufficient encouragement. One explanation might be that human intervention serves to transform chimpanzee conceptual space in a manner that cannot be achieved indigenously.

According to Boden's scheme, chimpanzees would be incapable of transforming their conceptual spaces for tool manufacture if they lack reflexive descriptions of their own tool-making procedures and ways of varying them. There is little doubt that chimpanzees possess knowledge of the properties of objects and the relationships between them (Byrne 1995). Nevertheless, it is unclear whether this implies reflexivity in the sense envisaged by Boden. It is true that chimpanzees do choose from a range of tools for specific purposes, but Byrne's observation that instances of genuine problem solving by great apes depend on 'noticing a solution when it comes by, not calculating it by some logical process' (ibid.: 85) suggests that true reflexivity is lacking. Byrne seems to be suggesting that, while chimpanzees can make new connections within the realm of their experience, they cannot think beyond it. Recast in Boden's terms, chimpanzee tool-use includes acts of improbabilist creativity but not impossibilist creativity. If Mithen (1996) is correct that early *Homo* possessed an enhanced technical intelligence, then perhaps this provided the kind of reflexivity that would enable the transformation of conceptual spaces relating to tool-use. If so, the manufacture of stone tools might provide evidence for impossibilist creativity in early *Homo*.

Even if stone tools do provide evidence for impossibilist creativity, there remains the issue of whether their appearance should be explained in terms of a few instances of local historical creativity or frequent episodes of personal psychological creativity. In other words, did early *Homo* live in a 'culture' of creativity? This is largely a question about the relative roles of individual and cultural learning in producing Oldowan technology. One can envisage at least two scenarios: one in which each individual learned for him or herself to manufacture stone tools, and another in which one or a few members of early *Homo* arrived at the idea of making stone tools and others then simply copied the new technology. The first scenario would imply the repeated exercise of a capacity for creative thought. The second is more problematic and depends upon the nature of the copying. Cognitive ethologists distinguish between several different types of copying, or social learning (Galef 1988; Whiten and Ham 1992). Programme-level imitation occurs when the novice understands the model's intention and acts to achieve the same end (Tomasello *et al.* 1993). Other types of social learning do not involve mind reading. For instance, stimulus enhancement occurs when the attention of the novice is drawn to some aspect of the environment by the model's behaviour (Whiten and Ham 1992). Cases of programme-level imitation require creative thought only if the novice attempts to achieve the model's end by a different means, whereas learning by stimulus enhancement always requires some degree of creativity.

The relationship between the different learning types and creativity can perhaps be summarised as follows. Individual learning of stone tool manufacture implies the exercise of both impossibilist and improbabilist creativity, on the grounds argued above. Copying, or social learning, does not require impossibilist learning, but depending on type, must or can require improbabilist learning. This is especially clear in the case of social facilitation, which functions to transform a conceptual space (by drawing attention to the possibility of working stone), but then requires exploration within the new space (learning to work stone). Thus in order to determine whether early *Homo* lived in a 'culture' of creativity (at least in the sphere of technology), we would ideally like to know how individual members of the genus learned to make tools.

Needless to say, we have no idea which members of a group actually made stone tools: all members or specialists, males, females or both? Nevertheless, the degree of variability in tool forms does perhaps provide some information about the type of learning involved. In principle, a high degree of variability would suggest a prevalence of individual learning, whereas the repetition of specific forms would suggest a prevalence of social learning. That having been said, three additional factors must be taken into account. Poor knapping skill or variable raw material quality would increase variability, while strong functional constraints might reduce it. The numerous step fractures on tools from Lokalalei suggest that the producers of the Omo industry were not accomplished knappers. In the case of the Oldowan, however, the evidence is more suggestive of a lack of design than poor technical skill. Toth (1985) has demonstrated by experiment that much of the variability in heavy- and light-duty tools can be attributed to the shape of the initial raw material. This argument is further supported by experimental (Jones 1981) and microwear (Keeley and Toth 1981) evidence for the general purpose nature of Oldowan technology: there is only limited correlation between specific tool forms and their apparent functions.

Since much of the variability in Oldowan tool forms can be attributed to the properties of the initial raw material, it cannot also be used to infer a lack of social learning. It does not follow, however, that aspects of technique were not socially learned just because specific tool *forms* were not transmitted through groups and across generations. Nevertheless, there does appear to be a real contrast with later tool industries in which the ubiquity of specific forms, such as the Acheulean handaxe, suggests a stronger component of social learning in tool manufacture (Isaac 1976). Indeed, Mithen (1994) has argued that the interdigitation of Acheulean and Oldowan type (Clactonian) industries in England can be attributed to changes in the intensity of social learning. He suggests that Acheulean industries are the result of strong social learning among members of large groups living in open habitats. In contrast, the variability *and relative lack of technical accomplishment* evident in the Clactonian industries is due to the prevalence of individual learning among members of small groups living in closed habitats.

If Mithen's model is broadly correct, then one might speculate that stone tool manufacture by early *Homo* was creative, as follows. To start with, there must have been one instance of historical impossibilist creativity that transformed the conceptual space of one individual to embrace stone as a workable raw material. Perhaps less trivially, there were also repeated instances of similar psychological creativity, but these were not frequent in the groups in which they occurred. Instead, most members of any given group learned to work stone as a result of social facilitation. If the latter serves to transform conceptual spaces, then for most individuals stone tool manufacture resulted from the exercise of improbabilist creativity within a socially transformed conceptual space.

CONCLUSION

According to Boden's scheme, the source of creativity ultimately lies with the thoughts of individuals. This is problematic for archaeologists dealing with remote periods of time because thoughts are simply not amenable to excavation. Nevertheless, it is equally true that we can never know for sure what many animals are thinking, yet the assumption that regularities in behaviour are referable to regularities in thought has proved to be a rewarding research strategy in cognitive ethology. Consequently, there is no reason, *a priori*, not to ask what hominids must have been thinking in order for them to have engaged in the behaviours for which there is archaeological evidence. However, Boden's scheme also requires an assessment of the limits of thought – a mapping of conceptual spaces. This proves an insurmountable obstacle in the attempt to establish whether early *Homo* exercised creativity in resource acquisition. Fortunately, the situation is not quite so hopeless in the case of stone tool manufacture, for which it is perhaps possible to offer some speculative suggestions. Specifically, it may be that the increased technical intelligence of early *Homo* facilitated reflexive descriptions of tool-making procedures, which in turn permitted the transformation of conceptual spaces to include stone as a workable material. One or more individuals must have exercised just this impossibilist creativity, but most probably never attained anything more than improbabilist creativity once their conceptual spaces had been transformed through social facilitation.

ACKNOWLEDGEMENTS

I am grateful to Steven Mithen for inviting me to contribute to this volume and for discussion of some of the issues surrounding the creativity of early *Homo*.

REFERENCES

Aiello, L. (1996) 'Hominine preadaptations for language and cognition.' In P. Mellars and K. Gibson (eds) *Modelling the Early Human Mind*, pp. 89–99. Cambridge: McDonald Institute for Archaeological Research.

Aiello, L. and Dunbar, R. (1993) 'Neocortex size, group size and the evolution of language.' *Current Anthropology* 34: 184–193.

Bilsborough, A. (1992) *Human Evolution*. Glasgow: Blackie Academic & Professional (Chapman & Hall).

Binford, L.R. (1981) *Bones: Ancient Men and Modern Myths*. New York: Academic Press.

Binford, L.R. (1983) *In Pursuit of the Past: Decoding the Archaeological Record*. London: Thames & Hudson.

Boden, M.A. (1994) 'Précis of *The Creative Mind: Myths and Mechanisms*.' *Behavioural and Brain Sciences* 17: 519–570.

Boesch, C. and Boesch, H. (1983) 'Optimisation of nut-cracking with natural hammers by wild chimpanzees.' *Behaviour* 83: 265–288.

Byrne, R. (1995) *The Thinking Ape: Evolutionary Origins of Intelligence*. Oxford: Oxford University Press.

Chamberlain, A.T. and Wood, B.A. (1987) 'Early hominid phylogeny.' *Journal of Human Evolution* 16: 119–133.

Chomsky, N. (1959) 'Review of Skinner (1957).' *Langauge* 35: 26–58.

Clark, J.D. (1970) *The Prehistory of Africa*. New York: Praeger.

de Waal, F. (1982) *Chimpanzee Politics*. London: Jonathan Cape.

Foley, R.A. (1987) *Another Unique Species*. Harlow: Longman Scientific & Technical.

Galef, B.G. (1988) 'Imitation in animals: history, definition and interpretation of data from the psychological laboratory.' In T.R. Zentall and B.G. Galef (eds) *Social Learning: A Comparative Approach*, pp. 3–28. Hillsdale, NJ: Erlbaum.

Gamble, C. (1993) *Timewalkers: The Prehistory of Global Colonization*. Stroud: Alan Sutton.

Grine, F. (1981) 'Trophic differences between "gracile" and "robust" australopithecines: a scanning electron microscope analysis of occlusal events.' *South African Journal of Science* 77: 203–230.

Groves, C.P. (1989) *A Theory of Human and Primate Evolution*. Oxford: Clarendon Press.

Hall, C.M., Walter, R.C. and York, D. (1985) 'Tuff above "Lucy" is over 3ma old.' *Eos* 66: 257.

Harris, J.W.K. and Johanson, D.C. (1983) 'Cultural beginnings: Plio-Pleistocene archaeological occurrences from the Afar, Ethiopia.' *The African Archaeological Review* 1: 3–31.

Isaac, G.L. (1971) 'The diet of early man: aspects of archaeological evidence from Lower and Middle Pleistocene sites in east Africa.' *World Archaeology* 2: 279–298.

Isaac, G.L. (1976) 'Stages of cultural elaboration in the Pleistocene: possible archaeological indicators of the development of language capabilities.' In S.R. Harnad, H.D. Steklis and J. Lancaster (eds) *Origins and Evolution of Language and Speech*, pp. 275–288. New York: Academy of Science.

Isaac, G.L. (1978) 'Food sharing and human evolution: archaeological evidence from the Plio-Pleistocene of east Africa.' *Journal of Anthropological Research* 34: 311–325.

Isaac, G.L. (1984) 'The archaeology of human origins: studies of the Lower Pleistocene in east Africa 1971–1981.' In F. Wendorf and A. Close (eds) *Advances in World Archaeology Volume 3*, pp. 1–79. New York: Academic Press.

Isaac, G.L., Harris, J.W.K. and Marshall, F. (1981) 'Small is informative: the application of the study of mini-sites and least effort criteria in the interpretation of the early Pleistocene archaeological record at Koobi Fora, Kenya.' In J.D. Clark and G.L. Isaac (eds) *Las Industrias mas Antiguas*, pp. 109–119. Mexico City: X Congresso, Union Internacional de Ciencias Prehistoricas y Protohistoricas.

Johanson, D.C. and Edy, M.A. (1980) *Lucy: The Beginnings of Human Kind*. New York: Simon & Schuster.

Johanson, D.C. and White, T.D. (1979) 'A systematic assessment of early African hominids.' *Science* 203: 321–330.

Johanson, D.C., Masao, F.T., Eck, G.G., White, T.D., Walter, R.C., Kimbel, W.H., Asfaw, B., Manega, P., Ndessokia, P. and Suwa, G. (1987) 'New partial skeleton of *Homo habilis* from Olduvai Gorge, Tanzania.' *Nature* 327: 205–209.

Jones, P.J. (1981) 'Experimental implement manufacture and use: a case study from Olduvai Gorge, Tanzania.' *Philosophical Transactions of the Royal Society of London Series B* 292: 189–195.

Jones, S., Martin, R. and Pilbeam, D. (eds) (1992) *The Cambridge Encyclopedia of Human Evolution*. Cambridge: Cambridge University Press.

Keeley, L.H. and Toth, N. (1981) 'Microwear polishes on early stone tools from Koobi Fora, Kenya.' *Nature* 293: 464–465.

Kibunjia, M. (1994) 'Pliocene archaeological occurrences in the Lake Turkana basin.' *Journal of Human Evolution* 27: 159–171.

Leakey, M.D. (1971) *Olduvai Gorge Volume 3: Excavations in Beds I and II 1960–1963*. Cambridge: Cambridge University Press.

McGrew, W.C. (1992) *Chimpanzee Material Culture: Implications for Human Evolution*. Cambridge: Cambridge University Press.

Mithen, S.J. (1991) 'Home bases and stone caches: the archaeology of early hominid activities (review article).' *Cambridge Archaeological Journal* 1: 277–283.

Mithen, S.J. (1994) 'Technology and society during the Middle Pleistocene.' *Cambridge Archaeological Journal* 4: 3–33.

Mithen, S.J. (1996) *The Prehistory of the Mind: A Search for the Origins of Art, Science and Religion*. London and New York: Thames & Hudson.

Potts, R. (1984) 'Home bases and early hominids.' *American Scientist* 72: 338–347.

Potts, R. (1988) *Early Hominid Activities at Olduvai Gorge*. New York: Aldine de Gruyter.

Potts, R. (1994) 'Variables versus models of Early Pleistocene hominid land use.' *Journal of Human Evolution* 27: 7–24.

Schick, K.D. and Toth, N. (1993) *Making Silent Stones Speak: Human Evolution and the Dawn of Technology*. New York: Simon & Schuster.

Semaw, S., Renne, P., Harris, J.W.K., Feibel, C.S., Bernor, R.L., Fesseha, N. and Mowbray, K. (1997) '2.5-million-year-old stone tools from Gona, Ethiopia.' *Nature* 385: 333–336

Shipman, P. and Harris, J.M. (1988) 'Habitat preference and palaeoecology of *Australopithecus boisei* in Eastern Africa.' In F.E. Grine (ed.) *Evolutionary History of the 'Robust' Australopithecines*, pp. 343–381. New York: Aldine de Gruyter.

Sibley, C. and Ahlquist, J. (1984) 'The phylogeny of hominoid primates as indicated by DNA–DNA hybridization.' *Journal of Molecular Evolution* 20: 2–15.

Skelton, R.R., McHenry, H.M. and Drawhorn, G.M. (1986) 'Phylogenetic analysis of early hominids.' *Current Anthropology* 27: 21–43.

Susman, R.L. (1991) 'Who made the Oldowan tools? Fossil evidence for tool behaviour in Plio-Pleistocene hominids.' *Journal of Anthropological Research* 47: 129–151.

Tomasello, M., Kruger, A.C. and Ratner, H.H. (1993) 'Cultural learning.' *Behavioural and Brain Sciences* 16: 495–552.

Toth, N.P. (1985) 'The Oldowan reassessed: a close look at early stone artefacts.' *Journal of Archaeological Science* 12: 101–120.

Toth, N.P., Schick, K.D., Savage Rumbaugh, E.S., Sevcik, R.A. and Rumbaugh, D.M. (1993) 'Pan the tool-maker: investigations into the stone tool-making and tool-using capabilities of a bonobo (*Pan paniscus*).' *Journal of Archaeological Science* 20: 81–91.

White, T.D., Suwa, G. and Asfaw, B. (1994) '*Australopithecus ramidus*, a new species of early hominid from Aramis, Ethiopia.' *Nature* 371: 306–312.

Whiten, A. and Byrne, R. (1988) 'The manipulation of attention in primate tactical deception.' In R. Byrne and A. Whiten (eds) *Machiavellian Intelligence: Social Expertise and the Evolution of Intellect in Monkeys, Apes, and Humans*, pp. 211–223. Oxford: Clarendon Press.

Whiten, A. and Ham, R. (1992) 'On the nature and evolution of imitation in the animal kingdom: reappraisal of a century of research.' *Advances in the Study of Behaviour* 21: 239–283.

WoldeGabriel, G., White, T.D., Suwa, G., Renne, P., Heinzelin, J. de, Hart, W.K. and Heiken, G. (1994) 'Ecological and temporal placement of Early Pleistocene hominids at Aramis, Ethiopia.' *Nature* 371: 330–333.

Wood, B.A. (1991) *Koobi Fora Research Project Vol. 4: Hominid Cranial Remains.* Oxford: Clarendon Press.

Wood, B.A. (1992a) 'Early hominid species and speciation.' *Journal of Human Evolution* 22: 351–365.

Wood, B.A. (1992b) 'Early *Homo*: how many species?' In L. Martin and W.H. Kimbel (eds) *Species, Species Concepts and Primate Evolution*, pp. 485–522. New York: Plenum.

Wood, B.A. (1997) 'The oldest whodunnit in the world.' *Nature* 385: 292–293.

CHAPTER NINE

MIDDLE PALAEOLITHIC 'CREATIVITY'

Reflections on an oxymoron?

STEVEN L. KUHN AND MARY C. STINER

When the editor of this volume first proposed the idea of writing a chapter on Neanderthal creativity, the initial reaction was that it would be a very short chapter indeed. The archaeological record of Neanderthals seems at first glance to provide little raw material for an essay on innovation. The Mousterian of Eurasia, the cultural period associated with the Neanderthals, is conspicuously bereft of evidence for artistic or aesthetic expression. Moreover, compared to later time periods, both artefacts and technology are remarkably uniform across space and stable over time during the Mousterian (Klein 1989: 296; Mellars 1989). Innovation is not usually the first word that comes to mind when one thinks of the Mousterian, but this makes the Mousterian all the more interesting from the perspective of the current volume. Attempting to account for the apparent absence of creativity in the material culture of the Neanderthals and other archaic hominids begs an examination of the general conditions that foster and encourage such creativity among later humans.

A number of recently published studies address possible changes in the structure of human/hominid cognition during the Pliocene and Pleistocene (Gibson and Ingold 1993; Mellars 1991; Mithen 1996a; Noble and Davidson 1996; Wynn 1989). Whereas the creative process is certainly a cognitive phenomenon, this chapter takes a somewhat different approach. Rather than addressing the mental foundations of innovation, we examine how unique creative acts can come to be visible in the archaeological record. Like many others, we assume that the very capacities of human beings for creative thought and action must have changed over the course of human evolution, and that this in turn must be reflected in the frequency with which novel objects and techniques appear in the archaeological record. However, it is one thing for a new and different idea to occur to an individual, but quite another for that idea to be actualised so as to produce the kind of durable and widespread material record that archaeologists can recognise. Of the two issues, the generation of a new idea and the repeated implementation or use of that idea, we focus on the latter. In light of the extremely 'coarse-grained' time resolution afforded by the Palaeolithic record, what is normally construed as evidence for creative behaviour – varied and rapidly changing material culture – may in fact reflect a more complex situation, consisting not only of the generation of novel designs or procedures but also of the widespread adoption and prolonged replication of these things. A diverse and dynamic archaeological record thus reflects both the creative tendencies of hominids and the ways in which creative behaviour was expressed, rewarded and disseminated within social groups.

Some aspects of the Palaeolithic record are more variable and show more rapid mutation than others, and, *a priori*, some aspects of material culture might be expected to be more dynamic than others. It is thus necessary to define where the best evidence for 'creativity' – and the most egregious lack of it – are situated within the Mousterian. Arguably, the most surprising lack of innovation, and the most pronounced contrast between the Middle Palaeolithic and later time ranges, is in the designs of stone artefacts, especially weapons. The record of modern humans leads us to expect a great deal of novelty and variety in this domain of technology, such that long periods of comparative stasis in the Middle Palaeolithic and earlier time ranges are truly noteworthy. Concentrating more narrowly on artefact design, it is possible to isolate the conditions which might foster long-term stability as well as the sorts of evolutionary factors which might lead to the apparent 'explosion' in creative expression during the later part of the Upper Pleistocene. At issue is the very role of technology in human adaptations: the extent to which material objects affect and are affected by other dimensions of human behaviour.

WHAT ARE WE TALKING ABOUT?

The Neanderthals are the most recent population of archaic *Homo* to have lived in Europe and southwest Asia. They are widely considered a subspecies of *H. sapiens* (*H.s. neanderthalensis*), although some researchers have revived the notion, prevalent up through the 1950s, that the Neanderthals represent a separate species (*H. neanderthalensis*). Archaeologically, the Neanderthals are most consistently associated with Mousterian or Middle Palaeolithic stone tool industries, a complex of archaeological assemblages found throughout western Eurasia. Although fossil remains classified as anatomically modern *H. sapiens sapiens* have also been found in association with Mousterian lithic assemblages at the sites of Skhul and Qafzeh in Israel (Vandermeersch 1981), the great majority of hominid remains from Mousterian sites are attributable to Neanderthals. There is no point in disputing the fossil associations, as they do show that skeletal 'modernity' is not equivalent to 'modern' behaviour. For the purpose of this discussion, we assume that the Eurasian Middle Palaeolithic is primarily but not exclusively the cultural record of the Neanderthals. Since this discussion concerns mainly archaeological patterns, the term 'Mousterian hominids' is more appropriate than 'Neanderthals', and the two designations are used more or less interchangeably below.

Mousterian assemblages are remarkably widespread, occurring from north Africa to the plains of southern Russia, and from the Atlantic coast of France to as far east as the Black Sea. Recent developments in chronometric dating have shown that the Middle Palaeolithic lasted for a very long time as well. The earliest assemblages conventionally classed as Mousterian date to in excess of 250,000 BP (e.g. Mercier *et al.* 1995), whereas the most recent are perhaps 33,000 years old (Delibrias and Fontugne 1990; Hublin *et al.* 1995). The range of artefact forms and debris that make up the Mousterian is relatively limited compared with later time periods. The surviving toolkit includes a variety of flake and blade tools, as well as bifaces, all of chipped stone. Bone was occasionally employed as a raw material, but only in an extremely casual, expedient manner (Vincent 1988). After chipped stone, animal bones, prey of both hominids and other predators, provide the next most abundant source of information about the behaviour of Neanderthals, although for historical reasons there have been few detailed studies of faunal remains until recently. The predominant prey species were medium to large herbivores (cervids, equids, bovids) (Mellars 1996: 193–244; Stiner 1993a). Smaller animals such as reptiles and shellfish were sometimes exploited as well (e.g. Stiner 1993b). So-called art objects and decorated utilitarian items attributable to Neanderthals or other Mousterian hominids are extremely rare, and usually controversial (Chase and Dibble 1987; Davidson and Noble 1989; Mellars 1996: 369–381).[1]

When referring to the Neanderthals as a taxonomic group, one is by definition discussing an entire hominid subspecies or perhaps even species. Similarly, the term Mousterian refers to archaeological remains representing an interval of around 200,000 years. While a mere instant of geological time, this is a substantial span of cultural time. The very fact that it is possible to encapsulate technology and subsistence of the Middle Palaeolithic and the Neanderthals in a few paragraphs is important. By contrast, it is difficult to imagine how one could summarise the material culture of *H. sapiens sapiens* over even the past 10,000 years in a comparable amount of text.

WHAT CONSTITUTES EVIDENCE FOR CREATIVITY, OR FOR ITS ABSENCE?

In discussing the Mousterian, it is important to decouple creativity in its broadest sense from the specifically expressive or 'artistic'. Although various authors have argued for the existence of objects with symbolic content dating as far back as the Middle Pleistocene (Bednarik 1995; Goren-Inbar 1986), these specimens are very rare. The fact that alleged Mousterian symbolic objects tend to be unique casts further doubt on their actual symbolic content (Chase and Dibble 1987; Davidson and Noble 1989), for they provide no evidence for a shared system of meaning. It is significant that Neanderthals as a subspecies so infrequently created symbolic objects of durable materials, but it also leaves little to work with. In addition, the frequency and elaboration of what archaeologists would identify as art varies widely among past and current groups of modern humans. Even within the late Upper Palaeolithic of Europe, thought by most to represent the first flowering of expressive art, the frequency and elaboration of art objects varies widely over time and across space. The rich parietal and mobile art traditions of southwest Europe dominate the public imagination, but they are in fact quite exceptional (Conkey 1983; Straus 1995). Because the proliferation of symbolic objects of durable materials seems often to occur under some social conditions but infrequently under others, it is difficult to know what exactly to expect of our more remote hominid predecessors.

It is also vital to consider the kinds of 'innovation' that might stand a reasonable chance of being detected in the Pleistocene archaeological record. Most of us tend to envision the creative act as unique – a novel idea that begins with a single individual. However, given the coarse chronological 'grain' of most Palaeolithic records, in which the finest possible subdivision may represent tens or even hundreds of years, individual acts and even individual lifetimes are for all intents and purposes invisible. A singular act of innovation will appear to the archaeological observer as part of a 'normal' range of variation for a particular time period or place. A truly radical innovation, expressed as an entirely novel type of artefact or technique, may not be recognised as such, or, if it anticipates later procedures or artefact forms, might be passed off as an intrusion of more recent material into the older level.

Given the nature of the record, the phenomena most likely to be considered evidence for innovation, or for the absence of it, are rates of turnover and degrees of variety in artefact forms, and not single objects or classes of things. While most would agree that rapid changes in what artefacts look like and how they are made indicate high rates of innovation, this is innovation on a very different scale. Palaeolithic archaeologists are most likely to recognise creativity only if its results are widely adopted and retained over long periods. Someone or something must first generate novelty, but in order for these developments to be recognised by observers thousands of years later, the novelty must have been widely disseminated and replicated. Ironically, it is only through the imitativeness, the lack of creativity of many, that we are likely to recognise the creativity of a few. The products of unique or uncommon acts will tend to be 'lost in the mix', archaeologically invisible. As a consequence, enquiring about the presence or absence of 'creativity' in the Palaeolithic, obliges us to consider the conditions that encourage individuals to innovate, as well as factors that impel others to adopt newly developed ways of doing things.

It is equally important to recognise that the static nature of the Middle Palaeolithic is obvious only by comparison with later time periods. Compared with the first million or so years of the Lower Palaeolithic, the Mousterian seems like a veritable Renaissance, an interval of constant fomentation. Technological constancy has been the rule in human evolution. The expectations that technological change should be rapid and ecologically responsive are based exclusively on experience with the record of modern humans over the past 40,000 years or so. It is actually the Upper Palaeolithic, or the record of later anatomically modern humans in general, that is anomalous. In light of these points, the issue is not what inhibited change in the earlier time ranges, but what it is about modern humans that makes their material culture so fluid.

WHERE IS THE ABSENCE OF INNOVATION MOST OBVIOUS?

Although the Mousterian/Middle Palaeolithic is often characterised as static and homogeneous, it does change, if incrementally, and it does vary geographically, if subtly (cf. Bar Yosef *et al.* 1992; Jelinek 1982; Mellars 1996: 315–355). Indeed, it is patently unrealistic to portray this or any other manifestation of human or hominid behaviour as totally fixed. Evolution cannot occur at all in the absence of variation, and Neanderthals could not have persisted for tens of thousands of years in some of the most extreme and unstable environments the world has ever seen without being quite flexible in their behaviour.

On closer analysis, it seems that the relative stasis and homogeneity that have impressed prehistorians actually reside in a few specific dimensions of the Middle Palaeolithic record. In some areas, such as in techniques for working stone, the Mousterian actually shows considerable diversity. A surprising variety of methods was employed for producing tool blanks, ranging from biface technology to prismatic blade production, and including a broad range of techniques labelled 'Levallois' (e.g. Boëda 1991, 1993a, 1993b; Boëda *et al.* 1990; Conard 1990; Marks and Monigal 1995; Mellars 1996: 56–94; Van Peer 1992). It could be argued that Mousterian lithic technology shows more variety than is manifest in Upper Palaeolithic stone tool production. Although distinct regional and chronological variants are present as well (e.g. Mellars 1996: 56–94), a significant component of the variation in methods of tool manufacture appears to represent responses to factors such as raw material forms and qualities, tool functions and even the logistics of keeping mobile individuals supplied with serviceable implements (e.g. Dibble 1985; Henry 1995; Jelinek 1988; Kuhn 1992, 1995; Rolland and Dibble 1990; Wengler 1990). A comparable degree of tactical flexibility appears to have characterised the procurement and transport of lithic raw materials (Geneste 1989a, 1989b; Roebroeks *et al.* 1988).

There is also good reason to believe that Middle Palaeolithic hominids exhibited a great deal of flexibility in their foraging and subsistence behaviour. The simple fact that the Mousterian is found over a vast area and across a broad range of palaeoenvironments demonstrates that the hominids of the time were capable of adjusting to the exigencies of making a living under radically different conditions. It is impossible, for example, that Neanderthal populations living in the southern Levant during hyperarid intervals pursued the same lifeways and exploited the same resources as did populations living in periglacial north-central Europe. When and where archaeologists have sought variability in Mousterian foraging economies over time or across space, they have quite often found it (e.g. Chase 1986; Farizy and David 1992; Mellars 1996: 193–244; Stiner 1993a, 1994). Whether Mousterian economic variability differs from what is observed in later periods (e.g. Klein 1989: 318–327; Stiner 1993a) is significant, but it is also a separate consideration.

A specific illustration of variation in Middle Palaeolithic technology and subsistence practices can be drawn from our research on several stratified Mousterian cave sites in west-central Italy (Kuhn 1995; Stiner 1994; Stiner and Kuhn 1992). In this area, distinct changes in foraging patterns and lithic technology occur around 55,000 BP, well within the temporal limits of the later Mousterian in Europe. Mousterian faunas dating to after this interval provide the first clear evidence for heavy dependence on hunting in the study area. Virtually entire carcasses of red deer, fallow deer and ibex were returned to the caves and eaten, often down to the marrow in the phalanges (Stiner 1994). In the most recent of these Mousterian faunas (from Grotta Breuil, see Bietti *et al.* 1990–91), foragers showed a marked tendency to target prime individuals of common ungulate species, a propensity rare among non-human predators but common in human populations during the Upper Pleistocene. This tendency persists in many cultures of the Holocene as well (Stiner 1990). The increasing emphasis on hunting in the Mousterian of west-central Italy is accompanied by a number of technological shifts, including declining numbers of exotic artefacts, markedly less intensive resharpening of tools, and changes in core reduction technologies, showing that cores of local raw materials, rather than more portable tools, were the primary object of conservation. Patterns of mobility provide the link between technological and faunal evidence. Briefly, before 55,000 BP, Mousterian subsistence in our study area was based to a large extent on dispersed small-package resources, including tortoises, shellfish and some scavenged game. This resulted in wide-ranging foraging patterns and a high degree of residential mobility, marked in the lithic assemblages by evidence for considerable reliance on extensively maintained transported tools. The provisioning of shelter sites with hunted game after 55,000 years ago is associated with somewhat longer stays in the caves, reducing dependence on transported toolkits, and allowing the Mousterians to collect small stores of scattered local raw materials that could be used and discarded as convenient (Kuhn 1992, 1995).

These strongly linked shifts in foraging and technology in this case reinforce the observation that all aspects of the Mousterian were not necessarily fixed or rigid. We also emphasise that most of the behavioural shifts documented in west-central Italy appear to have been responses to changing resource availability in and around some uniquely situated sites. They are evidence not of some profound evolutionary change in the capacities of the hominids, but of a simple adjustment to changing circumstances. The cave sites from which most of the archaeological data are drawn currently lie within 1 km of the Mediterranean sea, as they also did at the end of the last Interglacial. Declines in sea levels over the course of the later Pleistocene gradually exposed several kilometres of coastal plain, radically altering the nature of foraging opportunities available to Mousterian hominids using those shelters. Opportunities to hunt large grazing and browsing ungulates almost certainly improved as the sea receded and the environment around the caves became increasingly terrestrial in nature.

Returning to the central argument, the dimension of Middle Palaeolithic behaviour in which change is least noticeable is the designs of stone artefacts. What often strikes prehistorians familiar with later time periods is that the basic shapes of Mousterian tools seem to be virtually the same everywhere: very few edge types and tool forms comprise the bulk of all Middle Palaeolithic artefact assemblages. This homogeneity may be partially explained by the likelihood that most of Mousterian tools were used for processing food or working other materials (e.g. Anderson-Gerfaud 1990; Beyries 1987; Shea 1989), 'high tolerance' functions in which we might not expect a great deal of change or variety in artefact design. Progressive use and resharpening, the reduction effect (Dibble 1987) could further restrict the amount of variation in tool form that we observe. Yet, despite these qualifications, there remains a conspicuous absence of variety and novelty in Middle Palaeolithic artefact forms relative to later time periods.

The lack of variation in Mousterian tool design is perhaps most obvious in artefacts that might have been related to procuring and processing food, hunting weapons in particular. Given the vast geographic and ecological range over which Mousterian industries are found, it is inevitable that the degree to which people relied on large game animals varied extensively. Moreover, Mousterian hominids exploited many different species, ranging from small, semi-solitary ungulates such as gazelle and roe deer to large, gregarious animals like red deer, horses and wild cattle.

Relatively 'low-tolerance', time stressed activities such as the hunting of large animals place rather strict constraints on the design of implements (Bleed 1986; Torrence 1983), stimulating tool makers to develop a variety of more or less specialised forms. Yet potential Mousterian stone weapon tips, known as Levallois and Mousterian points, are made on very much the same plan wherever they occur. The main difference between the two forms is that Levallois points generally lack retouch, while Mousterian points are shaped by marginal flaking. Otherwise, both forms are large, essentially triangular stone flakes, sometimes with thinned butts. If hafted, both types of artefact are most likely to have been mounted on heavy thrusting or throwing spears (Churchill 1993; Shea 1993). Although the stone points do vary somewhat in size and elongation, differences in shape are generally attributable to raw materials or techniques of blank production.

The frequencies of likely projectile components vary across regions in unexpected ways as well. Potential weapon tips are not most abundant where one might expect on ecological grounds. For example, Mousterian and Levallois points tend to be more common relative to other 'tool forms' in assemblages from the southern end of the Mousterian range – the Levant and the Zagros mountains – rather than in more northerly areas where large animals must have played a more important role in the diet during cold seasons. It is not simply the case that Middle Palaeolithic foraging and technology were unrelated. In the Mousterian of west-central Italy, technology and foraging seem to have been linked mainly through land-use patterns, via their effects on lithic raw material economy. However, in our database, the shift from a notable reliance on scavenged game and small package resources like shellfish and tortoises to an emphasis on ambush hunting occurred around 55,000 years ago. Not long after this interval, and still well within the Mousterian, hunters were concentrating on prime-aged animals, the largest, fattest, but arguably the most difficult to obtain members of an ungulate population. Yet within the limits of available evidence, this important transition in human ecology was not accompanied by any substantial change in the lithic artefacts that might have been related to game procurement. Both the forms and the frequencies of potential stone weapon tips remained essentially constant across this entire interval in the study area.

In marked contrast to the Mousterian, the Upper Palaeolithic of Eurasia is characterised by a rich and rapidly changing array of weapons and other implements directly or indirectly related to food procurement. With the earliest Upper Palaeolithic, a remarkable variety of components for spears, darts, lances and harpoons make their appearance in Eurasia. This is not the place to develop a synthetic overview of potential Upper Palaeolithic weapons (see instead Knecht 1991, 1993; Larsen Peterkin 1993; Straus 1990, 1993). However, it is safe to say that within the comparatively brief span of the Upper Palaeolithic, both the di of implements for game procurement, and the variety of raw ma from which they were made, greatly exceed what even the most enthusiastic observer would identify for Middle Palaeolithic hunting technology (e.g. Shea 1989, 1993). Moreover, patterns of variation in the abundance and complexity of Upper Palaeolithic artefacts seem to be much more closely related to geographic and climatic factors than during earlier time periods, fitting well with general expectations about the importance of large game and other resources in human diets. For example, elaborate bone and antler projectiles, a defining characteristic of many later Upper Palaeolithic industries in northern Europe, are quite scarce in semi-arid southwest Asia during the later Upper Pleistocene. Conversely, ground and pecked stone tools apparently employed to process vegetable foods are much more common in the drier, warmer Near East during this same interval (e.g. Bar-Yosef and Belfer-Cohen 1988; Gilead 1991; Wright 1994).[2]

Here, then, lies one of the most ambiguous and fascinating aspects of the Middle Palaeolithic. On the one hand, it is obvious that the behaviour of Mousterian hominids was by no means rigid and unchanging. Neanderthals survived, even prospered, in a wide range of environments. They were more than capable of adjusting the ways in which they made stone tools to the diverse raw materials they encountered and even to the short-term tactical demands of keeping themselves supplied with usable tools in uncertain environments. Yet the designs of implements, a hallmark of innovative behaviour among modern humans and a particularly dynamic dimension of the later Upper Palaeolithic archaeological record, are curiously static for long periods of time within the Mousterian. Many of the functional and strategic factors that are expected to influence artefact design among later populations seem to have had little or no effect on how Neanderthals did things. We do not wish to argue that the absence of elaborate weapons technology compromised the predatory abilities of Neanderthals and their contemporaries. They were certainly capable of taking large game animals, and they probably used weapons to do so. What is strange is that there seems to be so little preserved evidence for technological responses to either documented or inferred shifts in how and how frequently large game was procured.

WHAT MIGHT LEAD TO ACCELERATED RATES OF CHANGE?

Although the persistence of Middle Palaeolithic artefact forms is striking, it is not without precedent: relative technological stasis has been the rule in human evolution for 2 million years or more. It is therefore unnecessary to invoke a special mechanism, such as exceedingly rigid and persistent cultural conventions, to explain why Neanderthals' stone tools changed so slowly. Rather than asking what kept the Neanderthals from changing, what we really should consider is what might have made the technologies of modern humans so dynamic by comparison with those of earlier hominids.

This is obviously an ambitious question, and we will not propose a comprehensive answer here. However, if one is going to ask why things change, specifically technology related to getting food, it is crucial to consider the conditions that would most stimulate rapid and constant change in the design of those tools. As discussed above, accounting for the kind of rapid turnover in artefact forms that an archaeologist might perceive as evidence for an acceleration in rates of technological innovation, in turn requires that we focus on both innovation and on the adoption and spread of novel forms, procedures and ideas.

To the extent that technology is a response to local conditions, instability and continuous change imply that there was almost always an advantage, roughly speaking, to getting better at something – in the case under discussion, the acquisition and preparation of food. Among modern humans, much of this pressure relates to patterns of sharing and the sizes of consumer groups. In foraging societies, food-sharing networks are (or were) usually quite flexible, almost open-ended, at least within the limits of local group sizes. No matter how much people collect or kill, there is someone to eat it, either immediately or in the future (with storage). Moreover, broad patterns of sharing appear to cement social ties and reinforce relationships in what may be called 'social storage'. A surprisingly broad cross-section of anthropologists, with theoretical backgrounds ranging from Marxism to sociobiology (see discussion in Hawkes 1992; Ingold 1991; Kelly 1995: 178–181), have noted how judicious redistribution of excess resources can provide a distinct advantage to some individuals, whether this advantage is measured in terms of increased survivorship of offspring, social prestige, reproductive opportunities, or some other currency. Under such conditions there may be, over the long term, a real benefit to becoming more efficient or faster at harvesting resources, even when they are not scarce in the environment relative to consumer demand

Long periods of relative stasis in the evolution of humans (and other organisms) show us that selective pressures on anatomy and behaviour were far from constant. Perhaps the monotony of early hominid food procurement technology – despite known or probable variation in diet – means that the potential payoffs for obtaining surpluses of food were not so open-ended among archaic humans as they are among modern peoples. The stability in archaic hominid food procurement technologies might reflect, albeit indirectly, something about the composition of social groups and the economic relationships between individuals within them. We do not mean to suggest that Neanderthals did not share food, or at least some kinds of food. It is well documented that large game were killed and carcasses sometimes transported to caves where they were extensively processed and consumed, implying either that game was regularly shared or else that individual Neanderthals were terrific gluttons. However, if sharing the fruits of foraging was nearly always limited to small groups – a female and her children, a mated pair and offspring, a group of close allies – then there would have been less general benefit to increasing the effectiveness of techniques for harvesting food resources in bulk so long as those resources remained at least moderately abundant in the environment.

It is instructive to look beyond a strictly human/hominid context with regard to the question of sharing as a stimulus for innovation in food procurement technology. In contrast to both humans and some social carnivores, non-human primates and most other omnivorous animals do not as a rule share food voluntarily. A good deal of what has been described as sharing among primates appears to be better described as 'tolerated theft', a situation in which the potential cost of defending a resource does not merit the effort (see Blurton-Jones 1987; King 1994: 65–67).[3] However, for pregnant or lactating females, a certain degree of sharing with offspring is obligatory, and demand for nourishment is as close to open-ended as it gets for an organism that doesn't habitually share. Interestingly, it seems that many 'advances' in primate foraging, particularly involving the use of tools or in techniques for processing food among chimpanzees, originate with or are most extensively exploited by females (e.g. Boesch and Boesch 1981, 1984; McGrew 1992: 88–106).

It is unlikely that a single model of group composition would serve for the entire Middle Palaeolithic: the actual sizes and compositions of Neanderthal groups probably varied somewhat, as would be expected for any social omnivore living under a comparably broad range of conditions. The point is that long-term stability of food procurement technologies during the Middle Palaeolithic (and earlier) may be in part attributable to the existence of sharing networks that were quite small in scale relative to what is known of modern foragers. Moreover, here the discussion concerns only the sizes and structures of economically cooperative, sharing social units. Other larger (and probably looser) social aggregates may well have existed, related to mating networks or defence against predators.

Although many of the developments in weapons technology during the Upper Palaeolithic do appear to have conferred tangible mechanical advantages, and presumably increased efficiency or effectiveness in procuring food, there is certainly more to it than that. It is unrealistic to think that all of the variation among and within human technologies reflects functional factors alone, and that every new form of implement that appears in the archaeological record is necessarily a better tool. It is abundantly clear today that the astounding variety in how people see fit to do and make things relates to much more than how those things actually function, and that the nature of a technology often has a great deal to do with the meaning assigned to technological acts in other domains of culture. Artefacts participate in the social and symbolic as well as the material realms of human existence. This very 'interconnectedness', the embedding of technology (or any other action) in the social, religious and economic domains, has consequences that might easily influence rates of technological change at an evolutionary time scale. Simply stated, the technologies of modern humans may be exceptionally volatile and diverse because they impact on so many aspects of cultural life, and because so many domains of culture impact on them. The multiplicity of influences on recent technologies may act almost like a mutagen acts on the genome, increasing variety on which selective processes may then operate.

In the diverse cultures of modern humans, an individual's proficiency as a maker of things and reputation as a holder of knowledge may have a profound impact on social position, the ability to attract mates, and access to other resources and information. There are sometimes great individual benefits to be gained by being – or appearing to be – a master of some aspect of technology. One way to demonstrate that mastery is to produce new and evidently better things. Moreover, because the constraints on artefacts' performance in the ideologic, social and functional domains are quite different, artefact design is simultaneously pushed in a variety of directions. Making an artefact work better as a medium for communicating social identity will not necessarily ensure that tools will be more effective in performing other practical tasks. On the other hand, this sort of process will encourage the production of novel forms, at least some of which may end up providing advantages to users in other domains. In some contexts, a premium on innovation or originality in design *per se* may exist, but this is not necessary to encourage elevated levels of technological creativity.

There are parallels between the evolution of somatic traits in animals and rates of change in artefact design. Anatomical features such as molar teeth, which are subject to stringent functional constraints, are well known to be highly conservative. On the other hand, features such as secondary sexual characteristics that are involved in competitive displays or physical conflicts designed to attract and gain access to mates are quite malleable on an evolutionary time scale. These features must still fill several roles: in addition to attracting a mate, a bird's feathers must also insulate and function in flight. However, the participation of a trait in more than one selective arena seems to confer a greater potential for rapid evolutionary change. In the case of features like pelage colour, feathers or antlers, a degree of novelty may in fact be favoured, as it causes an individual to stand out.

Of course, the value of a novel idea must be recognisable in order for others to emulate it. However, it is not always essential to build a significantly better mousetrap in order to have the world beat a path to your door. One can speculate that the mechanical characteristics of a particular innovation sometimes have relatively little impact on its initial spread and replication. Many developments in prehistoric technology seem to have provided such marginal functional advantages to the tool-user that it would be difficult for a single individual to evaluate them accurately over the short term. Initially, individuals may choose to adopt a novel way of doing things because it seems to work better, because of the reputation of the originator, or because it affords them some other benefit, such as identification with a particular subset of society. On the other hand, long-term retention and replication of a particular technological alternative, especially when it cross-cuts ecological and social boundaries, is more likely to reflect more fundamental, less contextually sensitive mechanical or functional factors.

Of course, not all aspects of modern technologies are equally dynamic. Some ways of doing or making things have been comparatively stable over long periods, while others change quite rapidly. Episodes of rapid diversification within our own society may occur when a technology begins to participate in a different domain of cultural life. The explosive growth in the athletic shoe industry in the United States is a case in point. For many years, the range of footgear available for sports and exercise was quite limited and relatively fixed; there were just a few brands, and each offered a few basic models. However, over the past 10–15 years, the range and variety in styles and 'functions' of athletic shoes has exploded, to the point that there are now large stores devoted exclusively to the sale of this one class of item. To be sure there have been changes in who wears athletic shoes and for what purpose, and there have certainly been advances in functionality. On the other hand, it is unlikely that the physical parameters of human exertion have changed enough in the last decade and a half to necessitate so many radical alterations in shoe design. Something else is afoot, namely that shoes have become social symbols as well as aids to athletic performance. Indeed, the athletic shoe as social symbol has spread well beyond the well-publicised cases of gang membership among inner city adolescents. Exercise, or at least exercise equipment, has become an important component of individual and social identity in many subsets of American society. In this instance, capitalist marketplace factors have amplified the scale of the response and the level of innovation in design, just as they have in other areas (the computer and car industries, for example). But the same basic phenomenon may characterise a wide range of human economic systems. The important point here is that the 'creative explosion' in athletic shoe design would never have occurred had the social valuation of sport not also changed.

This 'interconnectedness' between technology and other domains of human existence may be a unique characteristic of modern human culture beginning about 40,000 years ago. It does not seem to be an inevitable by-product of simply possessing some sort of culture and a minor dependence on tools. How much does a chimpanzee's skill at foraging or tool-use really impact on its opportunities for mating or its place in the social hierarchy? Probably very little, except insofar as the ability to obtain nourishment permits an individual to maintain a large body and a high level of activity. For a chimp, the activity of tool-using simply does not participate very much in any domain of life beyond the procuring of a few specific types of food.

The leisurely pace of change and broad spatial scale of variation in the technologies of Pleistocene human ancestors may imply an organisation of culture and a role of technology within it very different from what is typical of modern humans. If the various domains of early hominid cultures – especially the technological and social – were largely independent of one another, then the forces stimulating innovation and producing variety in technological behaviour, variety upon which selection could act, would have been much more limited. If all that Neanderthals did with tools was make other tools and help feed a small group of close allies or relatives, and if their knowledge of how to create things was of consequence only in this very limited arena, then the influences on technological behaviour would have been few, and the scope of innovation and change comparatively narrow. Under such conditions, we might not expect many aspects of hominid technology to change much more rapidly than hominid anatomy, which one can indeed argue was the case throughout the Lower and Middle Palaeolithic.

In turn, the acceleration in rates of technological change with the onset of the Upper Palaeolithic may be the equivalent of an 'adaptive radiation' in biological evolution, marking a point at which technology took on new roles in human existence. This expansion in the roles of technology seems to have occurred 60,000 years or more after the appearance of skeletally modern humans. For the first time in prehistory, artefacts began to have social and symbolic significance, and it is noteworthy that other media of communication such as body ornamentation first appeared and proliferated then (e.g. White 1989). Perhaps for the first time, tools and other artefacts took on the role of material culture as we now understand it. Whereas we certainly do not wish to argue that all changes in the forms of weapons or other tools, from the early Upper Palaeolithic on, reflect exclusively symbolic or social factors, once artefacts assumed 'functions' in these domains, the value of novelty and the rate at which new designs were produced and incorporated could have increased radically.

Why might technology have begun to participate in the social and ideological realms of prehistoric human life only at the beginning of the Upper Palaeolithic?[4] Clearly, the organisation of culture around symbolic interaction is a key. Symbols are, among other things, the currency that translates achievement in one domain into status in another. It is a common symbolic currency that qualifies a maker of fine leather clothing to be an oracle, that renders a good blacksmith a powerful magician, or that makes a successful hunter an interesting sexual partner. These are abstract, symbolic associations, much like the associations between a few lines of ochre on the wall of a cave and an actual horse. There are many strong opinions and little consensus about the symbolic capacities of Neanderthals and earlier hominids (see reviews in Mellars 1996: 388–391; Mithen 1996b; Noble and Davidson 1996); in fact, even the authors of this chapter do not agree about the extent to which Neanderthal communication resembled modern human language. However, regardless of what Neanderthals were capable of, there is certainly evidence for an amplification of many channels for information transmission with the beginning of the Upper Palaeolithic, whether in the appearance of well-defined and redundant 'art objects' or in abundant use of personal ornamentation (Gamble 1983; White 1989). By the beginning of the Upper Palaeolithic, human beings were clearly communicating with one another through material objects, if not for the first time, then far more than they ever had in the past. It is possible that these changes simply reflect that an organisational threshold had been reached, that regional populations had reached a level at which new channels of information transmission became necessary to alleviate conflict and establish boundaries. These same facts could also mark the first appearance of symbolic language as we now know it. We only comment that language is obviously central to, but not necessarily a precondition for, changes in both the sizes of human groups and the organisation of human cultures. In fact, the problems of avoiding conflict within ever denser regional populations, as well as of organising and coordinating action within increasingly large cooperating social groups (discussed by Gamble 1983; Whallon 1989; Wobst 1976, among others), could provide just the sort of selective context that pushed an already large-brained organism like *archaic H. sapiens* across a new threshold in communication.

CONCLUSION

Early hominid material culture, from the Oldowan through the Mousterian, is noticeably uniform and slow to change. In attempting to explain the slow pace of change in Middle Palaeolithic artefact designs, it is all too easy to assume that Neanderthal brains simply lacked the mental cog or sub-routine that provides the spark of creativity to modern humans. However, while Neanderthals probably did not think just like we do, they were equally large-brained, undeniably intelligent hominids who managed to flourish under an extraordinary range of conditions. Middle Palaeolithic behaviour was not, and indeed could not have been, completely static. Whether in foraging or in tool making, Mousterian hominids successfully coped with changing needs and conditions by altering what they did and how they did it. There was not necessarily anything stifling creativity or holding back change among Neanderthals and their forebears; it was simply that there was little spurring it on. The comparative dynamism of the Upper Palaeolithic, and of recent human material culture in general, stems at least in part from a major expansion in the role of technology in human culture. Sometime around 40,000 years ago (in Eurasia at least) it appears that material items came for the first time to participate regularly in the social and symbolic lives of humans (see also Mithen 1996a, 1996b). As an upshot of this 'radiation' of technology into new roles, the number of influences on the designs of implements increased exponentially, resulting in the kind of rapidly changing record that characterises the later Pleistocene in Eurasia and Africa. We emphasise that the specific social or ideological significance of particular material items is not at issue here. What is important is that material goods initially began to play a part in these domains of human existence in the Upper Palaeolithic.

A question to be addressed in future research is why there is evidence for the generation of novelty in some areas early on (techniques of stone working for example) and not in others, such as artefact design. Several possible resolutions to this problem present themselves. It is possible that innovation in methods of core reduction was actually favoured above creativity in artefact design, in that the factors influencing how stone is worked – raw material form and amiability, for example – required a more immediate response. On the other hand, the appearance of change in one domain and relative stasis in another may be simply a matter of what is available for comparison. As noted above, not all dimensions of the modern human archaeological record are equally dynamic and variable; for example, Late Holocene lithic technology is rather homogeneous across broad areas of the southwestern United States (Olszewski and Simmons 1982) at a time when ceramics diversified rapidly. It is possible that core reduction technologies in the Middle Palaeolithic seem so varied only because there was not so much variation during the later time periods. Perhaps because their utility as media for visual communication is limited, methods of flake and blade production never came to be of much wider social significance during the Upper Palaeolithic, so that they continued on the trajectory of very gradual change. It may also be the case that we simply know much more about variation in lithic technology during the Mousterian. Certainly, there has been more research on methods of blank production in the Middle and Lower Palaeolithic, and many of the more active researchers involved with the study of Palaeolithic *chaînes opératoires* are Mousterian specialists. Thus, while the surprisingly broad range of methods for flake production found in Mousterian assemblages does show that Middle Palaeolithic hominids did not always behave in a hidebound and stereotypical manner, in the end these phenomena may not actually be especially dynamic.

We have discussed Neanderthal 'creativity' at some length without referring explicitly to the cognitive abilities of those hominids. We have not addressed this topic because it lies outside our areas of greatest expertise, and because the cognitive aspects of the creative process are treated in depth by other contributors to this volume. In focusing mainly on the social contexts that might encourage technological innovation and the diffusion of novelty in material culture, we have attempted to draw attention to a more general set of factors that may be behind both long periods of stasis and intervals of explosive diversification in many aspects of human production and expression. It is very likely that fundamental shifts in human cognition accompanied the appearance of the Upper Palaeolithic. However, whether as prerequisite or consequence of changes in how hominids thought, conditions must have been right for cognitive developments to be expressed archaeologically. Moreover, while it is likely that the basic structure of human cognition has changed relatively little since the Upper Palaeolithic, the same principles behind the first 'radiations' in hominid technology during the Upper Palaeolithic continue to affect the world in which we live, albeit on a much more restricted scale.

NOTES

1 For extensive syntheses of the archaeology and physical anthropology of the Neanderthals, see Mellars (1996) and Stringer and Gamble (1993).
2 The focus of this chapter on implements that may have been used in procuring large animals reflects the composition of the archaeological record rather than some particular fascination with big-game hunting. Implements that might have been related to collecting and processing vegetable foods – milling, grinding, pounding tools – are extremely scarce and geographically unpatterned in the Middle Palaeolithic of western Eurasia.
3 Some primates do appear to share food regularly, but this accounts for a small portion of the total diet, and the sharing often takes place in a restricted social context (e.g. alliance formation among males) (e.g. Boesch and Boesch 1989; de Waal 1989; King 1994: 65–70; McGrew 1992: 106–113). Sharing is certainly not a ubiquitous organising principle in the foraging of any non-human primate, as it is among human beings (Hawkes 1992).
4 Whether the earliest symbolic objects and personal ornaments date to the very end of the Middle Palaeolithic or the beginning of the Upper Palaeolithic is not at issue here. A few cases of possibly symbolic objects, musical instruments or beads, found either in association with alleged Neanderthal fossils or dating to periods well before the appearance of anatomically modern humans in Europe, have been interpreted variously as evidence for the symbolising abilities of late Neanderthals or trade with early modern human groups. We are most concerned with the general timing of these developments and their broader evolutionary significance, less so with the species or subspecies involved.

REFERENCES

Anderson-Gerfaud, P. (1990) 'Aspects of behavior in the Middle Palaeolithic: functional analysis of stone tools from southwest France.' In P. Mellars (ed.) *The Emergence of Modern Humans: An Archaeological Perspective*, pp. 389–418. Edinburgh: Edinburgh University Press.

Bar Yosef, O. and Belfer-Cohen, A. (1988) 'The early Upper Paleolithic in Levantine caves.' In J. Hoffecker and C. Wolf (eds) *The Early Upper Paleolithic: Evidence from Europe and the Near East*, pp. 23–41. Oxford: British Archaeological Reports International Series 437.

Bar Yosef, O., Vandermeersch, B., Arensburg, B., Belfer-Cohen, A., Goldberg, P., Laville, H., Meignen, L., Rak, Y., Speth, J., Tchernov, E., Tillier, A.-M. and Weiner, S. (1992) 'The excavations in Kebara Cave, Mt. Carmel.' *Current Anthropology* 33: 497–550.

Bednarik, R. (1995) 'Concept-mediated marking in the Lower Paleolithic.' *Current Anthropology* 36: 605–634.

Beyries, S. (1987) *Variabilité de l'Industrie Lithique au Moustérien: Approche Fonctionelle sur quelques Gisments Françaises*. Oxford: British Archaeological Reports International Series S328.

Bietti, A., Kuhn, S., Segre, A. and Stiner, M. (1990–91) 'Grotta Breuil: introduction and stratigraphy.' *Quaternaria Nova* 1: 305–325.

Bleed, P. (1986) 'The optimal design of hunting weapons.' *American Antiquity* 51: 737–747.

Blurton-Jones, N. (1987) 'Tolerated theft, suggestions about the ecology and evolution of sharing, hoarding, and scrounging.' *Ethology and Sociobiology* 11: 353–359.

Boëda, E. (1991) 'Approche de la variabilité des systemes de production lithique des industries du Paléolithique inférieur et moyen: chronique d'une variabilité attendue.' *Techniques et Culture* 17–18: 37–79.

Boëda, E. (1993a) 'Le débitage discoïde et le débitage Levallois recurrente centripête.' *B.S.P.F.* 90: 392–404.

Boëda, E. (1993b) *Le Concept Levallois: Variabilité des Méthodes*. Paris: C.N.R.S.

Boëda, E., Geneste, J.-M., and Meignen, L. (1990) 'Identification de chaînes opératoires lithiques du Paléolithique ancien et moyen.' *Paleo* 2: 43–80.

Boesch, C. and Boesch, H. (1981) 'Sex differences in the use of natural hammers by wild chimpanzees: a preliminary report.' *Journal of Human Evolution* 10: 585–593.

Boesch, C. and Boesch, H. (1984) 'Possible causes of sex differences in the use of natural hammers by wild chimpanzees.' *Journal of Human Evolution* 13: 415–440.

Boesch, C. and Boesch, H. (1989) 'Hunting behavior of wild chimpanzees in the Taï National Park.' *American Journal of Physical Anthropology* 78: 547–573.

Chase, P. (1986) *The Hunters of Combe Grenal: Approaches to Middle Paleolithic Subsistence in Europe*. Oxford: British Archaeological Reports International Series 286.

Chase, P. and Dibble, H. (1987) 'Middle Paleolithic symbolism: a review of current evidence and interpretations.' *Journal of Anthropological Archaeology* 10: 193–214.

Churchill, S. (1993) 'Weapon technology, prey size selection, and hunting methods in modern hunter-gatherers: implications for hunting in the Paleolithic and Mesolithic.' In G. Peterkin, H. Bricker and P. Mellars (eds) *Hunting and Animal Exploitation in the Later Paleolithic and Mesolithic of Eurasia*, pp. 11–24. Archaeological Papers of the American Anthropological Association, 4.

Conard, N. (1990) 'Laminar lithic assemblages from the last interglacial complex in northwestern Europe.' *Journal of Anthropological Research* 46: 243–262.

Conkey, M. (1983) 'On the origins of Paleolithic art: a review and some critical thoughts.' In E. Trinkaus (ed.) *The Mousterian Legacy: Human Biocultural Change in the Upper Pleistocene*, pp. 201–227. Oxford: British Archaeological Reports International Series 64.

Davidson, I. and Noble, W. (1989) 'The archaeology of perception: traces of depiction and language.' *Current Anthropology* 30: 125–155.

Delibrias, G. and Fontugne, M. (1990) 'Datations des gisments de l'Aurignacien et du Moustérien en France.' In C. Farizy (ed.) *Paléolithique Moyen Récent et Paléolithique Supérieur Ancien en Europe*, pp. 39–42. Nemours: Mémoirs du Musée de Préhistoire d'Ile de France, n. 3.

de Waal, F. (1989) 'Food sharing and reciprocal obligations among chimpanzees.' *Journal of Human Evolution* 18: 433–459.

Dibble, H. (1985) 'Raw material variation in Levallois flake manufacture.' *Current Anthropology* 26: 391–393.

Dibble, H. (1987) 'The interpretation of Middle Paleolithic scraper morphology.' *American Antiquity* 52: 109–117.

Farizy, C. and David, F. (1992) 'Subsistence and behavioral patterns of some Middle Paleolithic local groups.' In H. Dibble and P. Mellars (eds) *The Middle Paleolithic: Adaptation, Behavior, and Variability*, pp. 85–96. University Museum Monographs No. 72. Philadelphia: University of Pennsylvania.

Gamble, C. (1983) 'Culture and society in the Upper Palaeolithic of Europe'. In G. Bailey (ed.) *Hunter-Gatherer Economy in Prehistory: A European Perspective*, pp. 201–211. Cambridge: Cambridge University Press.

Geneste, J.-M. (1989a) 'Economie des ressources lithiques dans le Moustérien du sud-ouest de la France.' In M. Otte (ed.) *L'Homme de Néandertal, Vol. 6: La Subsistence*, pp. 75–97. Liège: ERAUL 33.

Geneste, J.-M. (1989b) 'Systèmes d'approvisionnement en matières premières au Paléolithique moyen et au Paléolithique supérieur en Aquitaine.' In M. Otte (ed.) *L'Homme de Néandertal, Vol. 8: La Technique*, pp 61–70. Liège: ERAUL 35.

Gibson, K. and Ingold, T. (1993) *Tools, Language and Cognition in Human Evolution*. Cambridge: Cambridge University Press.

Gilead, I. (1991) 'The Upper Paleolithic period in the Levant.' *Journal of World Prehistory* 5: 105–153.

Goren-Inbar, N. (1986) 'A figurine from the Acheulean site of Berekhat Ram.' *Mi'tekufat Ha'even* 19: 7–12.

Hawkes, K. (1992) 'Sharing and collective action.' In E. Smith and B. Winterhalder (eds) *Evolutionary Ecology and Human Behavior*, pp. 269–300. New York: Aldine de Gruyter.

Henry, D. (1995) 'The influence of mobility levels on Levallois point production, late Levantine Mousterian, southern Jordan.' In H. Dibble and O. Bar-Yosef (eds) *The Definition and Interpretation of Levallois Technology*, pp. 185–200. Madison, WI: Prehistory Press.

Hublin, J., Ruiz, C., Lara, P., Fontugne, M. and Reyss, J. (1995) 'The Mousterian site of Zafarraya (Andalucia, Spain). Dating and implications for the Paleolithic peopling processes of western Europe [in French].' *Comptes Rendues de l'Academie des Sciences*, Serie II, fasc. A, *Sciences de la Terre et des Palenetes* 321: 931–932.

Ingold, T. (1991) 'Notes on the foraging mode of production.' In T. Ingold,

D. Riches and J. Woodburn (eds) *Hunters and Gatherers 1: History, Evolution, and Social Change*, pp. 269–285. New York: Berg.

Jelinek, A. (1982) 'The Middle Paleolithic in the southern Levant, with comments on the appearance of modern *Homo sapiens*.' In A. Ronen (ed.) *The Transition from Lower to Middle Paleolithic and the Origin of Modern Man*, pp. 57–101. Oxford: British Archaeological Reports International Series 151.

Jelinek, A. (1988) 'Technology, typology, and culture in the Middle Paleolithic.' In A. Montet-White and H. Dibble (eds) *Upper Pleistocene Prehistory of Western Eurasia*. pp. 199–212. University Museum Monographs No. 54. Philadelphia: University of Pennsylvania.

Kelly, R. (1995) *The Foraging Spectrum: Diversity in Hunter-Gatherer Lifeways*. Washington: Smithsonian Institution Press.

King, B. (1994) *The Information Continuum. Evolution of Social Information Transfer in Monkeys, Apes, and Hominids*. Santa Fe: School of American Research Press.

Klein, R.G. (1989) *The Human Career: Human Biological and Cultural Origins*. Chicago: University of Chicago Press.

Knecht, H. (1991) *Technological Innovation and Design During the Early Upper Paleolithic: A Study of Organic Projectile Technologies*. Ph.D. Dissertation, Department of Anthropology, New York University.

Knecht, H. (1993) 'Early Upper Paleolithic approaches to bone and antler projectile technology.' In G. Peterkin, H. Bricker and P. Mellars (eds) *Hunting and Animal Exploitation in the Later Paleolithic and Mesolithic of Eurasia*, pp. 33–48. Archaeological Papers of the American Anthropological Association, 4.

Kuhn, S. (1991) '"Unpacking reduction": lithic raw material economy in the Mousterian of west-central Italy.' *Journal of Anthropological Archaeology* 10: 76–106.

Kuhn, S. (1992) 'On planning and curated technologies in the Middle Paleolithic.' *Journal of Anthropological Research* 48: 185–214.

Kuhn, S. (1995) *Mousterian Lithic Technology: An Ecological Approach*. Princeton, NJ: Princeton University Press.

Larsen Peterkin, G. (1993) 'Lithic and organic hunting technology in the French Upper Paleolithic.' In G. Peterkin, H. Bricker and P. Mellars (eds) *Hunting and Animal Exploitation in the Later Paleolithic and Mesolithic of Eurasia*, pp. 49–88. Archaeological Papers of the American Anthropological Association, 4.

McGrew, W. (1992) *Chimpanzee Material Culture: Implications for Human Evolution*. Cambridge: Cambridge University Press.

Marks, A. and Monigal, K. (1995) 'Modeling the production of elongated blanks from the early Levantine Mousterian at Rosh ein Mor.' In H. Dibble and O. Bar-Yosef (eds) *The Definition and Interpretation of Levallois Technology*, pp. 267–278. Madison, WI: Prehistory Press.

Mellars, P. (1989) 'Technological changes across the Middle–Upper Paleolithic transition: technological, social, and cognitive perspectives.' In P. Mellars and C. Stringer (eds) *The Human Revolution: Behavioral and Biological Perspectives on the Origins of Modern Humans*, pp. 338–365. Princeton, NJ: Princeton University Press.

Mellars, P. (1991) 'Cognitive changes and the emergence of modern humans.' *Cambridge Archaeological Journal* 1: 63–76.

Mellars, P. (1996) *The Neandertal Legacy*. Princeton, NJ: Princeton University Press.

Mercier, N., Valladas, H., Valladas, G., Reyss, J.-L., Jelinek, A., Meignen, L. and Joron, J.-L. (1995) 'TL dates of burnt flints from Jelinek's excavations at Tabun and their implications.' *Journal of Archaeological Science* 22: 495–509.

Mithen, S. (1996a) *The Prehistory of the Mind*. London and New York: Thames & Hudson.

Mithen, S. (1996b) 'Social learning and cultural tradition: interpreting Early Palaeolithic technology.' In J. Steele and S. Shennan (eds) *The Archaeology of Human Ancestry: Power, Sex and Tradition*, pp. 207–229. London: Routledge.

Noble, W. and Davidson, I. (1996) *Human Evolution, Language and Mind*. Cambridge: Cambridge University Press.

Olszewski, D. and Simmons, A. (1982) 'Tools for thought: some comments on the analysis of Puebloan chipped stone assemblages.' *The Kiva* 48: 109–116.

Roebroeks, W., Kolen, J. and Rensink, E. (1988) 'Planning depth, anticipation, and the organization of Middle Paleolithic technology: the archaic natives meet Eve's descendants.' *Helinium* 28: 17–34.

Rolland, N. and Dibble, H. (1990) 'A new synthesis of Middle Paleolithic variability.' *American Antiquity* 55: 480–499.

Shea, J. (1989) 'A functional study of the lithic industries associated with hominid fossils in Kebara and Qafzeh cave, Israel.' In P. Mellars and C. Stringer (eds) *The Human Revolution: Behavioral and Biological Perspectives on the Origins of Modern Humans*, pp. 611–625. Princeton, NJ: Princeton University Press.

Shea, J. (1993) 'Lithic use-wear evidence for hunting by Neandertals and early modern humans from the Levantine Mousterian.' In G. Peterkin, H. Bricker and P. Mellars (eds) *Hunting and Animal Exploitation in the Later Paleolithic and Mesolithic of Eurasia*, pp. 189–197. Archaeological Papers of the American Anthropological Association, 4.

Stiner, M. (1990) 'The use of mortality patterns in archaeological studies of hominid predatory adaptations.' *Journal of Anthropological Archaeology* 9: 305–351.

Stiner, M. (1993a) 'Modern human origins – faunal perspectives.' *Annual Review of Anthropology* 22: 55–82.

Stiner, M. (1993b) 'Small animal exploitation and its relation to hunting, scavenging, and gathering in the Italian Mousterian.' In G. Peterkin, H. Bricker and P. Mellars (eds) *Hunting and Animal Exploitation in the Later Paleolithic and Mesolithic of Eurasia*, pp. 107–126. Archaeological Papers of the American Anthropological Association, 4.

Stiner, M. (1994) *Honor among Thieves: A Zooarchaeological Study of Neanderthal Ecology*. Princeton, NJ: Princeton University Press.

Stiner, M. and Kuhn, S. (1992) 'Subsistence, technology, and adaptive variation in Middle Paleolithic Italy.' *American Anthropologist* 94: 306–339.

Straus, L. (1990) 'The original arms race: Iberian perspectives on the Solutrean phenomenon.' In J. Kozlowski (ed.) *Feuilles de Pierre: Les Industries à Pointes Foliacées du Paléolithique Supérieur Européen*, pp. 425–447. Liège: ERAUL 42.

Straus, L. (1993) 'Upper Paleolithic hunting tactics and weapons in western Europe.' In G. Peterkin, H. Bricker and P. Mellars (eds) *Hunting and Animal Exploitation in the later Paleolithic and Mesolithic of Eurasia*, pp. 83–94. Archaeological Papers of the American Anthropological Association, 4.

Straus, L. (1995) 'The Upper Paleolithic of Europe: an overview.' *Evolutionary Anthropology* 4: 4–16.

Stringer, C. and Gamble, C. (1993) *In Search of the Neanderthals: Solving the Puzzle of Human Origins*. London: Thames & Hudson.

Torrence, R. (1983) 'Time budgeting and hunter-gatherer technology.' In G. Bailey (ed.) *Hunter-Gatherer Economy in Prehistory: A European Perspective*, pp. 11–22. Cambridge: Cambridge University Press.

Vandermeersch, B. (1981) *Les Hommes Fossiles de Qafzeh (Israël)*. Paris: C.N.R.S.

Van Peer, P. (1992) *The Levallois Reduction Strategy*. Madison, WI: Prehistory Press.

Vincent, A. (1988) 'L'os comme artifact au Paléothique moyen: Principes d'étude et premiers résultats.' In M. Otte (ed.) *L'Homme de Néandertal, Vol. 4: La Technique*, pp. 185–196. Liège: ERAUL 31.

Wengler, L. (1990) 'Économie des matières premières et territoires dans le Moustérien et l'Atérien maghrébiens. Exemples du Maroc oriental.' *L'Anthropologie* 94: 335–360.

Whallon, R. (1989) 'Elements of culture change in the Later Paleolithic.' In P. Mellars and C. Stringer (eds) *The Human Revolution: Behavioral and Biological Perspectives on the Origins of Modern Humans*, pp. 433–454. Princeton, NJ: Princeton University Press.

White, R. (1989) 'Production complexity and standardization in early Aurignacian bead and pendant manufacture: evolutionary implications.' In P. Mellars and C. Stringer (eds) *The Human Revolution: Behavioral and Biological Perspectives on the Origins of Modern Humans*, pp. 366–390. Princeton, NJ: Princeton University Press.

Wobst, M. (1976) 'Locational relationships in Paleolithic society.' *Journal of Human Evolution* 5: 49–58.

Wright, K. (1994) 'Ground-stone tools and hunter-gatherer subsistence in southwest Asia: Implications for the transition to farming.' *American Antiquity* 59: 238–263.

Wynn, T. (1989) *The Evolution of Spatial Competence*. Urbana, IL: Urbana University Press.

CHAPTER TEN

A CREATIVE EXPLOSION?

Theory of mind, language and the disembodied mind of the Upper Palaeolithic

STEVEN MITHEN

Art makes a dramatic appearance in the archaeological record. For over 2.5 million years after the first stone tools appear, the closest we get to art are a few scratches on unshaped pieces of bone and stone (Bednarik 1992, 1995). It is possible that these scratches have symbolic significance – but this is highly unlikely (Chase and Dibble 1987, 1992; Davidson 1992; Mithen 1996a). They may not even be intentionally made. And then, a mere 30,000 years ago, at least 70,000 years after the appearance of anatomically modern humans, we find cave paintings in southwest France – paintings that are technically masterful and full of emotive power. Their appearance is one of the changes in the archaeological record that mark the start of the Upper Palaeolithic in Europe. The apparent sudden appearance of art may be no more than an artefact of the processes of preservation and discovery (Bednarik 1994). It is possible that images were being created in non-durable media for many thousands of years prior to the painting of these French caves. It is also possible that works of art await to be discovered in the archaeological record from much earlier times. Indeed, artefacts from the Early Palaeolithic (*c*.2.5 mya to 50,000/30,000 years ago) that are claimed to have symbolic significance, or even to have representational status, are occasionally published within the archaeological literature – such as the highly ambiguous Berekhat Ram 'figurine' (Marshack 1997; Goren-Inbar 1986; Pelcin 1994). The status of claimed 'art' in Australia at Jinmium dating to *c*.75,000 BP (Fullagger *et al.* 1996) remains unclear with regard to both the reliability and significance of the dates associated with the images, and the character of the art itself.

The apparent sudden appearance of art in the archaeological record has in fact become more firmly established during the last few years. The dating of cave paintings and engravings has always been contentious.

Chronological schemes based on stylistic evolution pegged on to the circumstantial dating of a few selected images were developed by Breuil and Leroi-Gourhan (see review in Bahn and Vertut 1988: ch. 4). But these had their own internal contradictions (Clottes 1990), and it has only been with the advent of AMS radiocarbon dating that a firm, uncontentious chronology has begun to be established (Valladas *et al.* 1992). This dating method allows minute samples of charcoal to be removed from the pigment itself and has shown several paintings to be many thousands of years older than had been expected. The earliest dates are found at Chauvet cave in the Ardèche region of France, where an image of a rhino has been dated to 32,417 ± 720 radiocarbon years BP. Three other paintings in the cave have a mean date of approximately 30,000 years ago, while painting at Cosquer cave has a date of 27,110 ± 390, and those in Cougnac and Pech Merle caves date prior to 20,000 years (Chauvet *et al.* 1996).

These earliest works of art are believed to be exclusively associated with anatomically modern humans. Although Neanderthals were present in Europe when Chauvet was painted, the continuity of the Upper Palaeolithic art tradition through to 10,000 years ago, 20,000 years after the Neanderthals had become extinct, excludes the possibility that they had been responsible for painting Chauvet. Indeed, all of the cultural developments after 50,000 years ago appear to have been created by anatomically modern humans, although Neanderthals may have been attempting to mimic some of the new types of artefacts (Mellars 1989a).

The earliest paintings show all the technical sophistication and emotive power of the more well-known paintings from Lascaux and Altamira – those paintings that are universally recognised as one of the great creative achievements of human culture. The paintings and carvings are dominated by animals, often depicted with a high degree of naturalism (e.g. Figure 10.1) and displaying considerable anatomical knowledge (Chauvet *et al.* 1996). At Chauvet, impressions of herds are created by the superimpositioning of paintings, and a higher frequency of carnivores are depicted than is generally found with the Upper Palaeolithic painted caves (Clottes 1996). Unreal creatures are also found within this art, composites of two or more species, together with abstract signs and stencils of human hands. Chauvet itself has been compared to Lascaux in terms of the brilliance of its art – even though these are the very first paintings known to humankind.

While we must always remain open to the possibility of new discoveries, the fact that the appearance of these paintings broadly coincides

Figure 10.1 Head of a bison carved on a 'baton' from Isturitz, Pyrenées-Atlantiques. Head c.5 cm

Source: Drawn by K. Bambridge after Leroi-Gourhan 1965

with a host of other new types of behaviour, such as body adornment (White 1989), new technology (Mellars 1989a) and the colonisation of arid areas (Gamble 1993), can only strengthen our belief that they do signify a major transition in the nature of human thought and behaviour at this very late stage in human evolution. These other developments are not all simultaneous. They appear to emerge piecemeal, initially with the burials in the Near East 100,000–80,000 years ago – the first to include gravegoods – and then with the colonisation of Australia 60,000 years ago which required substantial sea crossings (Allen 1994; Davidson and Noble 1992). The first art appears in Europe at 30,000 years ago, and in other continents soon after that date (see review in Mithen 1996b). So there appears to be a window of cultural development beginning 100,000 years ago that reaches a crescendo at 30,000 years ago with what we can recognise as fully modern behaviour. These changes in the archaeological record and human behaviour, which in Europe are referred to as the transition from the Middle to the Upper Palaeolithic, and in Africa as that from the Middle to the Late Stone Age, have been described and discussed at immense length during the past three decades (e.g. Bar-Yosef 1994; Klein 1989; Mellars 1973, 1989a; Mithen 1996b; Soffer 1994; Stringer and Gamble 1993; White 1982).

It is common within that literature for this transition to be described as a creative explosion, following Pfeiffer's (1982) use of this to title his book about the period. It would indeed seem perverse not to do so. For it is with this transition that we see the beginning of those types of behaviour for which the word creative is most readily applied – the production of art and technological invention. And two central features of 'creativity' are certainly applicable: we see immense *novelty* in the archaeological record, and that novelty is clearly *valued*, as new ideas, inventions and ways of behaving appear to spread extremely rapidly across vast distances and become a permanent feature of human society.

In this chapter I wish to address the emergence of a new degree of creative thought at this late stage of human evolution. To do so, I will initially summarise arguments made at greater length elsewhere (Mithen 1996b) regarding the change from a domain-specific to a cognitively fluid mentality, and explain how these relate to Boden's (1990) explanation of creative thought. I will then expand on these arguments by elaborating on three aspects of behaviour and mind that underlie the 'creative explosion': the possession of a 'theory of mind', the evolution of language, and the role of material culture as a non-biological extension of mind.

CREATIVE THOUGHT AND THE EXPLORATION OF CONCEPTUAL SPACES

Margaret Boden (1990, 1994) has argued that creative thought can be explained 'in terms of the mapping, exploration, and transformation of structured conceptual spaces'. Her definition of conceptual spaces is vague: she describes them as a 'style of thinking – in music, sculpture, choreography, chemistry, etc.'. In spite of this vagueness, the idea of transforming conceptual spaces is intuitively appealing. It has a close association with the earlier notions of Koestler (1964), who has described creative thinking as arising from the sudden interlocking of two previously unrelated skills or matrices of thought, and the contemporary ideas of Perkins (1994), who uses the terminology of 'klondike spaces' and argues that these are often systematically explored in the process of creative thinking. In this regard, while creative thinking is clearly part of ordinary thinking (Hodder, this volume; Weisberg 1993), and not something restricted to 'geniuses', we can nevertheless see the potential for how particularly creative thoughts may arise from quite unusual transformations of conceptual spaces undertaken by particular individuals in particular circumstances.

Boden (1990) draws support for her explanation of creative thought from studies of child development. In particular she draws on the work of Karmiloff-Smith, who examined how the drawing skills of children develop with age and reflect, according to Boden, an ever increasing ability to map their own conceptual spaces. Karmiloff-Smith's work shows how 4-year-olds have an inflexible 'man drawing procedure', making it difficult, perhaps impossible, for them to draw imaginary men, such as those with two heads. At a slightly older age, drawing skills are 'mapped as a list of distinct parts which can be individually repeated and rearranged in various ways' (Boden 1990: 71). This enables 'funny' pictures to be drawn, such as men with extra arms. But the degree of flexibility remains limited – imaginary beings remain largely absent. Ten year olds, however, produce a vast range of man pictures, including images with multiple heads, limbs or bodies. They display a far greater ability to map, explore and transform their conceptual spaces.

While Boden drew on such research to support her general characterisation of creativity, she did not place her ideas into an evolutionary context. Yet, as I have argued in detail elsewhere (Mithen 1993, 1996b), the emergence of an ability to map, transform and explore conceptual spaces appears to be precisely what we witness at the Middle/Upper Palaeolithic transition.

Humans before this period of dramatic cultural change, whether they were anatomically modern humans, Neanderthals, *archaic H. sapiens*, *H. erectus*, or any other of the species now identified in the fossil record, appear to have possessed a domain-specific mentality (Mithen 1996b, 1996c). By this I mean that their ways of thinking and stores of knowledge about different cognitive domains, including those of stone tools, the natural world and social interaction, were effectively isolated from each other. Rather than using the terminology of 'conceptual spaces', I referred to these as 'cognitive domains' – but the idea has significant similarities to those of Boden. With regard to pre-modern humans, there may have been an ability to explore each of their conceptual spaces/cognitive domains on an individual basis, but not to transform them by integrating stores of knowledge or ways of thought from different spaces/domains. The 'creative explosion' of the Upper Palaeolithic was a result, I have argued, of a newly evolved ability for precisely the 'mapping, exploration and transformation' of conceptual spaces that Boden has identified as lying at the core of creative thought (Figure 10.2). I have termed this as a capacity for 'cognitive fluidity' and suggested that this is a universal feature of modern minds, differentiating us from Early Humans, and is the root cause of our ability to engage in creative thinking.

Boden (1990, this volume) describes some of the ways in which transformations of conceptual spaces might arise. One of these is by dropping a constraint, such that a composer might do to introduce novel features into his or her music. An ability to combine knowledge from multiple cognitive domains provides the potential to drop constraints on a grand scale. For instance, Early Humans with a domain-specific mentality are likely to have believed that any living entity will have required food to survive – that will have been part of the natural history intelligence (Mithen 1996b). By bringing ideas about inert objects into contact with those beliefs, may enable them to drop this constraint, resulting in ideas about living entities that can survive without food, indeed entities that do not obey any biological rules at all – such as ghosts and spirits, which are very general, perhaps universal, features of religious beliefs (Boyer 1994). Early Humans may have been able to 'drop constraints' or undertake other methods of transforming conceptual spaces that Boden describes, such as considering negatives, but this will have been possible only within – not across – cognitive domains. Hence, without cognitive fluidity the overall potential for creative thought is markedly constrained.

The critical question is how did this 'cognitive fluidity', this ability to map, explore and transform conceptual spaces, arise in evolutionary time? To answer this, I now want to elaborate previous suggestions I have made regarding this issue. To do so, I will examine what appear to me to be the three critical evolutionary developments in the mind that underlie the emergence of cognitive fluidity and which explain that burst of creative activity in the Upper Palaeolithic: theory of mind, modern language and the disembodiment of the mind via material culture.

THEORY OF MIND AND CREATIVE THOUGHT

One of the most exciting areas of current research in the cognitive sciences is that concerning 'theory of mind'. This refers to an ability to attribute a full range of mental states to other individuals as well as oneself, and then to use such attributions to predict and understand behaviour. A vast literature has arisen regarding this issue, as it seems to

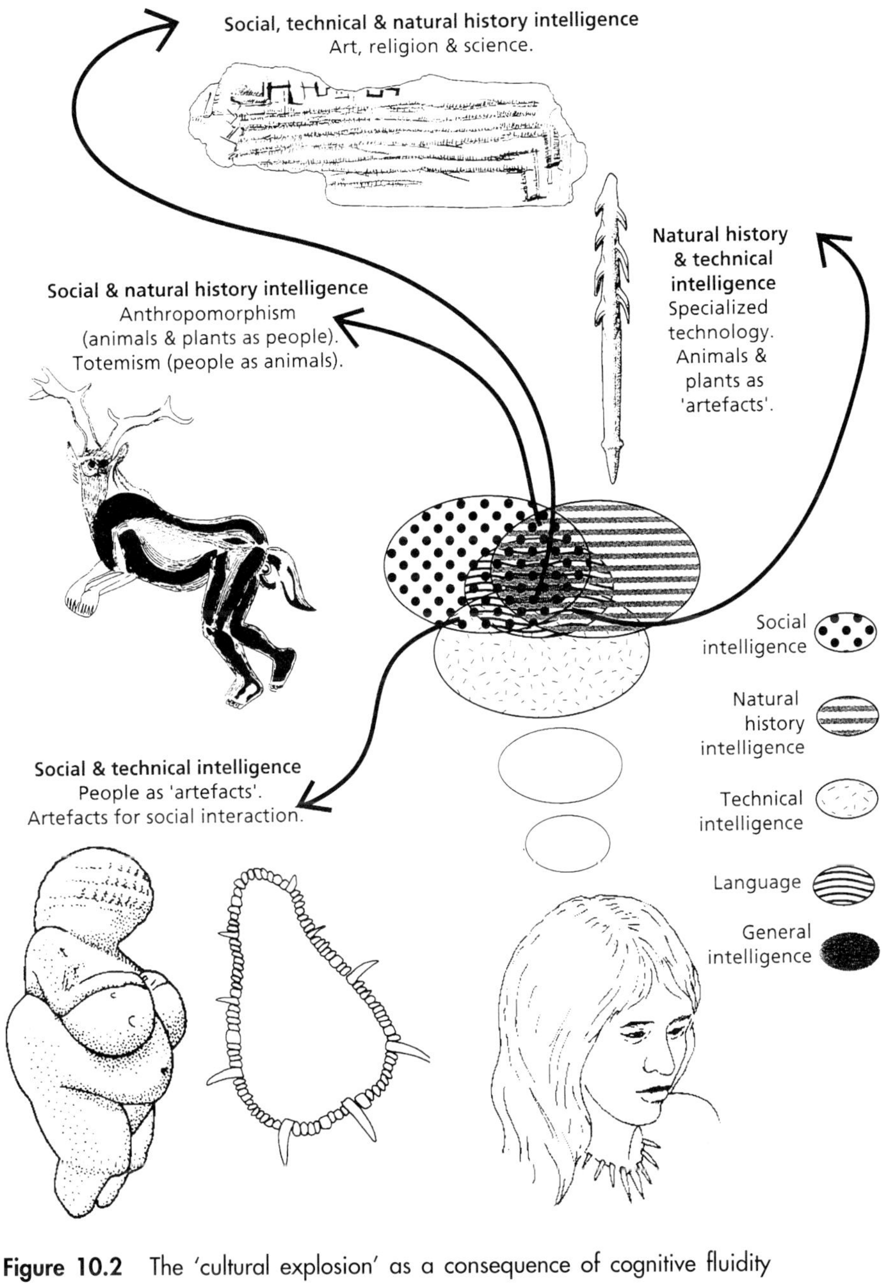

Figure 10.2 The 'cultural explosion' as a consequence of cognitive fluidity
Source: Mithen 1996b

be one of the most critical cognitive features humans possess. As such, theory of mind is a central element of what has been termed 'social' and 'Machiavellian intelligence' (Byrne and Whiten 1988) and is closely related to, perhaps synonymous with, 'mind reading' (Whiten 1991), 'folk psychology', 'natural psychology' (Humphrey 1976) and the 'intentional stance' (Dennett 1988).

The importance of theory of mind for human thought and behaviour has been stressed by Baron-Cohen (in press). He lists nine behaviours that depend on the possession of a theory of mind:

1 Intentionally informing others
2 Intentionally deceiving others
3 Intentionally communicating with others
4 Repairing failed communication with others
5 Teaching others
6 Intentionally persuading others
7 Building shared plans and goals
8 Intentionally sharing a focus or topic of attention
9 Pretending

Without these behaviours, human society would be very different. How different might be appreciated by considering that current research strongly suggests that a dysfunctioning of the part of the brain that enables theory of mind may be the root cause of autism (Baron-Cohen 1995). Autistic children do not appear to engage in any of the above behaviours – their lack of a theory of mind has a devastating effect on their abilities to socialise and communicate. It also appears to inhibit their imagination, which suggests that the possession of a theory of mind may provide a critical foundation for creative thinking. From the above list of behaviours, pretence is that which is most clearly linked to imagination and creative thinking. Pretending requires one to hold in one's mind the actual identity of an object, along with the pretend identity. As such, one must have knowledge about one's own mind. If that is lacking, so too will be the ability to pretend.

The dependency of creative thinking on theory of mind is also apparent from a recent set of drawing experiments that makes a direct link between the experiments by Karmiloff-Smith, described above, and the imagery in the earliest of the Upper Palaeolithic art. This link is the ability to produce representations of unreal things. In Upper Palaeolithic art, imaginary beasts play a significant role. The most dramatic example is the lion/man from Hohlenstein-Stadel, Germany (Hahn 1993). This is a *c.*11-cm-high figure carved from a single mammoth tusk about 30,000 years ago, which has the body of a man and the head of a lion. A second impressive example is the bison/man painting from Chauvet cave, which has the head and shoulders of a bison and human legs (Chauvet *et al.* 1996). Other paintings within that cave date to 30,000 years ago, and appear to be part of the same early Upper Palaeolithic artistic tradition as the figure from Hohlenstein-Stadel. Imaginary animals, 'monsters' and composite figures are found throughout the Upper Palaeolithic art tradition, with a notable collection coming from the cave of Pergouset, assumed to be Magdalenian in date (15–12,000 BP, Lorblanchet and Sieveking 1997) and further examples from Les Trois Frères (Figure 10.3), Pech Merle and Gabillou (Leroi-Gourhan 1983).

A link to the research on the theory of mind from these images is that autistic children appear unable to draw imaginary animals. Scott and Baron-Cohen (1996) undertook a series of experiments that demonstrated that autistic children have a deficit in the representation of unreal things, as had been argued by Leslie (1991). They appear unable to produce just those type of images that we find in the earliest Upper Palaeolithic art, and which we intuitively feel are more creative than those other images of real animals and people. Two explanations were proposed for this deficit. The first is that those suffering from autism are unable to 'fuse' together two primary representations of real objects. Scott and Baron-Cohen give the example of representing a 'flying pig', which involves joining together representations of a real pig, and a real bird, to produce a representation of an unreal animal, the flying pig. An alternative idea relates more directly to an absence of a theory of mind. They argue that 'representing an unreal object necessarily requires pretending, or representing that you (the agent) are holding a pretend *attitude* (or *mental state*) towards an object' (Scott and Baron-Cohen 1996: 381; italics in original).

A second deficit in those suffering from autism appears to be an inability to engage in pretend play (Baron-Cohen 1995). One might also draw links here with the new types of behaviour that emerge with the Upper Palaeolithic. Pretence is a critical feature of ritual in extant hunter-gatherer societies. There can be little doubt that ritualised behaviour was also a critical feature of the earliest Upper Palaeolithic societies. Some of the composite figures in the art seem most likely to be a shaman in costume. That from Trois Frères (Figure 10.3), for example, has the legs and posture of a man, the back and ears of a feline, the antlers of a deer and a phallus positioned like that of a feline (Bahn and Vertut 1988).

In summary, the possession of a theory of mind appears to be a critical requirement for imaginative play and creative thinking; it appears to be an essential prerequisite for the exploration and mapping of conceptual spaces, even if quite why and how are not fully understood. One possibility, therefore, is that the creative explosion of the Upper Palaeolithic is directly related to the evolution of the theory of mind. Does the absence of representations of imaginary animals and ritualised activities prior to 30,000 years ago suggest that a theory of mind was also absent?

The evolution of 'theory of mind'

This appears to be unlikely for two reasons. The first is that evidence exists for a theory of mind among our closest living relatives, the chimpanzees and gorillas. We shared a common ancestor with these species no more than 6 mya, and if they possess a theory of mind, then we must conclude that a theory of mind was also present in our own earliest ancestors. Byrne (1995, this volume) has summarised the experimental and observational evidence for a theory of mind among chimpanzees, of which deceptive behaviour is the most important, and is confident that mental state attribution does exist (although this appears absent in monkeys). The major criticism of this evidence is that it is largely anecdotal (Heyes 1993a). Nevertheless, I find the material accumulated by Byrne sufficient to indicate that chimpanzees possess a form of theory of mind, admittedly one that may not be as well developed as that of modern humans.

As some degree of theory of mind exists in chimpanzees, and so by implication in our common ancestor of 6 mya, we are unable to attribute the creative explosion of 35,000 years to the emergence of this capacity. This is further supported by the substantial expansion of the brain that occurred after 2.5 mya (Aiello and Dunbar 1993). It seems most unlikely that this did not result in some elaboration of the theory of mind capacity – what else was the processing power of the enlarged brain being used for? Robin Dunbar (1993, 1996) has argued that the large brain of our human ancestors implies living in social groups substantially larger than those of existing non-human primates. It is clear from many recent studies that these non-human primates needed to deploy a Machiavellian intelligence to maintain the balance between group fusion and fission and hence maintain an appropriate group size for the efficient exploitation of foodstuffs and predator defence (Byrne and Whiten 1988; Humphrey 1976). Theory of mind is most likely to be an essential feature of this Machiavellian intelligence and would have become increasingly important if group size increased during the Middle Pleistocene in the manner that Dunbar claims. In other words, theory of mind appears to have been an essential prerequisite for social life of Early Humans.

One further strand of evidence to support this concerns the evolution of language. As I will summarise below, there is strong anatomical evidence that by 250,000 at least, Early Humans had anatomical adaptations for speech. This most likely implies intentional communication, which as Baron-Cohen (in press) argues, requires a theory of mind.

To summarise: if creativity is about the exploration and transformation of conceptual spaces, this appears to have been possible only with the existence of a theory of mind. But that appears to have arisen in human evolution much earlier than the creative explosion of the Upper Palaeolithic. Consequently, we must now turn to the issue of language, and its relationship to creative thought.

LANGUAGE AND CREATIVE THOUGHT

Just as it seems perverse not to describe the cultural developmen 60–30,000 years ago as a 'creative explosion', it seems impossi believe that these are associated with a change in the nature of language (Corballis 1992; Davidson and Noble 1989; Mellars 1991; Whallon 1989). Just as art, science and religion are unique to *H. sapiens sapiens* in the modern world, so too is language. If creativity is simply an element of 'ordinary thinking' (Wiesberg 1993), and if theory of mind is a necessary but not sufficient condition, then we must consider the role of language in human thought, in the exploration of conceptual spaces and in the creative explosion of the Upper Palaeolithic.

When did language evolve?

First, we must consider when language evolved. For this, the fossil and archaeological records provide contradictory evidence. The fossilised remains of our human ancestors indicate that language appeared early in human evolution, being dependent upon a series of anatomical pre-adaptations (Aiello 1996a, 1996b). Evidence for this is found on brain endocasts that indicate that Broca's area, conventionally associated with speech, was already present in *H. habilis*, *c*.2 mya (Falk 1983; Tobias 1987), well developed in *H. erectus* by 1.5 mya (Begun and Walker 1993), and essentially modern on *H. neanderthalensis* (Holloway 1981a, 1981b, 1985). Evidence for a sophisticated speech capacity is also found in terms of the Neanderthal vocal tract, which is now believed to be not significantly different from that of modern humans (Houghton 1993; Schepartz 1993). This was made particularly evident by the discovery of a Neanderthal hyoid bone in 1983, which had a form virtually identical to that of modern humans (Arensburg *et al.* 1990). A further source of evidence for the presence of language relatively early in human evolution might be found in the pattern of encaphalisation. It is now clear that modern brain size had arisen between 600,000 and 150,000 years ago (Ruff *et al.* 1997). At present, there is no reason necessarily to expect that modern brain size implies a language capacity. But Dunbar (1993; Aiello and Dunbar 1993) provides a powerful argument that such a relationship may exist. In summary, the fossil evidence strongly suggests that, at least from 500,000 years ago, our ancestors had sophisticated capacities for speech. Whether this should be described as language or 'proto-language' is an issue I will discuss below.

This fossil evidence for the evolution of spoken language appears to be contradicted, however, by that from the archaeological record. This seems to provide no trace of language mediated behaviour until very late in human evolution (Davidson 1991). The most obvious would be visual symbols – language after all is a system of audible symbols and it may seem unlikely, to say the least, that symbolic capacities might exist in one medium but not the other. Bednarik (1995) and Marshack (1990) claim that visual symbols do exist and point to artefacts such as scratched bones from Bilzingsleben (Figure 10.4) or the so-called Berekhat Ram figure from Israel. This latter piece is a fragment of volcanic rock that is claimed to have been intentionally modified into a female figurine (Marshack 1997). Although I find Pelcin's (1994) explanation of the marks as having a natural source more convincing, even if one concludes that this and other early artefacts are intentionally incised, this does not necessarily imply a symbolic capacity on a continuum with our own. I think we should compare these artefacts to true visual symbols in the same manner as we compare the 'language' of apes to the language of humans (Mithen 1996a). Ape language is likely to be derived from different cognitive processes from those used by humans to generate their utterances. Those used by apes are likely to be part of generalised learning abilities rather than deriving from dedicated language learning modules. Similarly, if non-utilitarian, intentionally incised artefacts prior to 50,000 years ago do indeed exist, and they are functioning as visual symbols, then I suspect they arose from generalised learning abilities rather than the type of cognitive processes that underlie the true symbolism of modern humans.

Another trace of language-mediated behaviour that we might expect to see in light of the fossil evidence for language is the distribution of knapping waste in such a manner that implies an apprentice was being taught by a craftsman. Such distributions are certainly present in the later Upper Palaeolithic (e.g. Pigeot 1990) but absent from early prehistory, in spite of excellently preserved knapping floors, such as at Boxgrove (Roberts 1986). Some might argue, however, that simply the presence of biface and Levallois technology amongst Early Humans suggests language. Could the transmission of the technical skill to make such artefacts have been possible without spoken language? Wynn (1991) thinks it could have been and supports this by explaining how, in modern society, speech is rarely used to transmit technical skill.

A further possible type of language-mediated behaviour among Early Humans is their subsistence behaviour, especially hunting, for which cooperation appears likely. Evidence for hunting by Neanderthals comes from sites dominated by single prey species (e.g. the bovids at Mauran; Girard and David 1982) and particular patterns of cutmarks on bones, such as at Combe Grenal (Chase 1986). Conclusive evidence that even earlier humans were big-game hunters has come from the discovery of spears at Schöningen (Thieme 1997) dating to 400,000 years ago, likely to be associated with *H. heidelbergensis*. But the cognitive implications of big-game hunting remains unclear. We see, for instance, a range of social carnivores as well as chimpanzees hunting very effectively together in teams with no evident communication (Boesch and Boesch 1989). Moreover, while big-game hunting can no longer be doubted to have existed

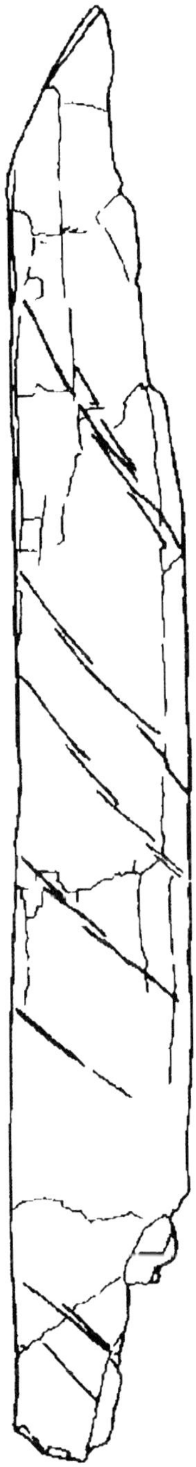

Figure 10.3 Fragment of a rib of a large mammal from Bilzingsleben, Germany, possibly engraved with a series of parallel lines, 28.6 cm

Source: Redrawn from Mania and Mania 1988

among Early Humans, the extent and character of this hunting remains to be established. Our lack of knowledge is clearly evident from the range of interpretations that exist regarding the La Cotte mammoth remains. This stacked pile of bones within a cave at the base of a cliff may have ultimately arisen from animals being driven off the edge of a cliff. But whether this was carefully planned group hunting (Scott 1980) or 'dangerous driving by desperate men' (Stringer and Gamble 1993: 162) remains unclear.

To summarise: although the fossil evidence clearly indicates the presence of language, we see tenuous, if any, reflection of this in the traces of behaviour preserved in the archaeological record. How can this paradox be resolved? The most obvious solution is that the type of language possessed by Early Humans was quite unlike language today. Two alternative proposals have been put forward for the nature of this 'proto-language', that by Bickerton (1990, 1996) and that by Dunbar (1993, 1996).

Non-grammatical proto-language

The most well-developed argument about proto-language has been formulated by the linguist Derek Bickerton (1996). For him, proto-language lacked syntax. It included a vast lexicon, but the grammatical rules for organising those words were extremely simple; indeed he has even speculated about a 'random ordering of words'. He argues that syntax evolved as a catastrophic event, transforming the nature of language and, as a consequence, the mind. To do so, the neural networks that are likely to form the basis of syntax are assumed to have been partly or wholly in place early in human evolution, and Bickerton acknowledges that these required a respectable amount of evolutionary time in which to appear. But he believes that during their evolution they played a role in the mind/brain quite different from that of syntax, which emerged following a final mutation *c.*50,000 years ago.

Bickerton believes that language without syntax results in much poorer abilities not only to communicate, but also to think. He makes a distinction between on-line and off-line thinking. The former refers to 'computations carried out only in terms of neural responses elicited by the presence of external objects, while off-line thinking involves computations carried out on more lasting internal representations of those objects' (Bickerton 1996: 90). Not surprisingly, Bickerton believes that off-line thinking is unique to humans, is dependent upon language with its complex syntax, and lies at the heart of our particular form of intelligence and consciousness. On-line thinking is still a vital ingredient of the modern mind, but just as language has been added on to an ancestral communication, so too has off-line thinking been added to the mind. Whereas non-linguistic minds can have a primary representation of entities, only when syntactical language is present are neural workspaces available for secondary, more abstract representations. Only with off-line thinking can our thoughts be released from the exigencies of the moment and structured into complex wholes.

This theory would appear to be compatible with the archaeological record. It implies that even though Early Humans may have been using speech and language, we should not expect to see many traces of language-mediated behaviour, and certainly no traces of abstract thought, for such language would have been essentially unrelated to thought. But there are several problems with the theory. First is the supposedly catastrophic emergence of syntax. If these neural networks that underlie syntax did indeed already exist in the brain, then Bickerton needs to provide some suggestion as to what they were being used for and how they could so easily be changed to fulfil a new role. Moreover, it remains unclear as to how the presence of syntax results in off-line thinking. Perhaps this is indeed simply the freeing of 'brain power', because speech and language become dramatically more efficient. In this scenario, the large brain of Early Humans had been selected to process the vast number of non-syntactical utterances within proto-language. Carruthers (in press) has suggested that non-syntactical proto-language would have placed great demands on the on-line processing, interpretation and storage of utterances. Once grammar had evolved, interpretation of utterances would have become semi-automatic, thus freeing cognitive space for other tasks. My own feeling is that although Bickerton is correct that a form of proto-language preceded modern language, and that this was principally used for communication rather than thought, he is wrong to believe that proto-language differed in terms of structure by lacking syntax. A much more convincing argument is that it differed in content.

Proto-language as social language

This is the argument from Robin Dunbar (1993, 1996) and Leslie Aiello (Aiello and Dunbar 1993), who argue that language evolved as a 'social language'. The general purpose functions of language, its use in all domains of behaviour, arose, they argue, only at a relatively late stage in human evolution. The basis of their argument is that as group size enlarged during the Middle Pleistocene, social language was selected as a mechanism for exchanging social information. When group size was relatively small, social grooming would have been sufficient to exchange adequate social information. But as group size enlarged during the Pleistocene, the need for a greater amount of social information increased exponentially, providing the selective pressures for new forms of communication to supplement, and then replace, grooming, i.e. language. The claimed enlargement of group size during the Pleistocene is both the most critical and contentious part of their theory. They base their argument on a correlation between brain size and group size that exists among extant non-human primates. As Early Humans had much larger brains than any living primate, they are also assumed to have lived in much larger group sizes, estimated to be *c.*250 for Neanderthals. But this, of course, involves a substantial extrapolation of the relationship between brain size and group size, the validity of which may be questioned.

Dunbar and Aiello provide little information about the nature of this supposed social language, but one need not exclude the possibility of a complex syntax being present. The critical difference from modern language is that it was not being used to communicate about non-social domains of activity, such as tool making or foraging. As such, this idea is quite compatible with the fossil and archaeological evidence – perhaps even more so than the theories of Bickerton. If language was restricted to the social domain, we should not be surprised to find a lack of evidence for language-mediated behaviour in the archaeological record, which effectively consists of the remains of foraging and tool-making activities. Prior to the Upper Palaeolithic, these do appear not to have become embedded into the social domain as they are among modern humans (Mithen 1996b).

In this scenario for proto-language, the creative explosion of the Upper Palaeolithic is intimately related to the transition from a social to a general purpose language. As it adopted its general purpose function, it provided the mechanism by which the 'barriers' between cognitive domains were broken down (Mithen 1996b: 185–194).

Language and a 'community mind'

It was by this change in the nature of language, therefore, that the ability to explore, map and transform one's conceptual spaces became possible. In fact, language does much more than this, for it allows conceptual spaces to be created and explored that are no longer part of a single mind/brain. Language connects people's minds into a network, allowing ideas to 'migrate' between minds, exponentially increasing the number of conceptual spaces, and the extent to which they can be explored. Andy Clark captures this remarkable power of language:

> migrations [of ideas by language] may allow the communal construction of extremely delicate and difficult intellectual trajectories and progressions. An idea that only Joe's prior experience could make available, but that can flourish only in the intellectual niche currently provided by the brain of Mary, can now realise its full potential by journeying between Joe and Mary as and when required. The path to a good idea can now criss-cross individual learning histories so that one agent's local minimum becomes another's potent building block . . . culturally scaffolded reason is able to incrementally explore spaces which path dependent individual reason could never hope to penetrate.
>
> (Clark 1996: 206)

So we see that the evolution of modern, general purpose language played a dual role in forming the modern mind and delivering the potential for creative thought. It provided the means by which one could explore one's own conceptual spaces, and, by creating a network of minds, the extent of this exploration and transformation was exponentially increased. As Clark describes, language is a means by which mind is extended beyond the bounds of individual brains and bodies. With language, the mind becomes disembodied. And this leads us directly to the role of material culture and creative thought.

MATERIAL CULTURE AND CREATIVE THOUGHT

We traditionally assume that the extraordinary elaboration in material culture that we witness at the start of the Upper Palaeolithic in Europe requires explanation as a *product* of cognitive or behaviour changes. It is more appropriate to view this material culture as much a cause as a product of such changes, in terms of a positive feedback loop that generated a transformation in human mind, behaviour and culture within a short period of time – the creative explosion.

Material culture acted in this manner as it performs a role similar to that of language in terms of creating networks of minds, disembodying minds, and exponentially increasing the range of conceptual spaces available for exploration and the manner in which this could be undertaken. The quotation from Clark given above can apply to material culture as well as language; in fact, it is even more appropriate for material culture. While language is effective at allowing ideas to migrate between minds, it has a major constraint – utterances are not durable. If there is no one around to hear what one says, the idea within the utterance is lost, or at least remains within the mind of the speaker. But once ideas are encoded into durable media, they become part of the world and can be communicated across vast time spans – as archaeologists are only too aware each time they pick up an ancient stone tool or visit a painted cave. As such, material culture is the prime means by which minds are extended out of the body, and connected between individuals who may never meet each other, or even know of each other's existence.

There are three aspects of material culture to consider in this regard: first, material culture as a means of the storage of information; second, material culture as a means to anchor ideas that have no evolutionary basis within the mind; and third, the requirement for constant reinterpretations of material culture and the ideas that it may encode. I will consider each in turn.

Material culture as non-biological memory

Archaeologists have long recognised that much of the 'artwork' of the Upper Palaeolithic may be functioning to store information, epitomised in Alexander Marshack's (1972, 1991) suggestions that certain incised artefacts may have been lunar calendars. Yet it was a psychologist, Merlin Donald (1991), who explored this theme in greatest detail by suggesting that these material objects were a form of 'external symbolic storage': in effect, they constituted an extension of biological memory. Francesco D'Errico (1991, 1995; D'Errico and Cacho 1994) has further developed this theme by attempting to formulate a theory of external memory aids and developing a rigorous methodology for their identification. The significance of 'external memory' may not be immediately recognised by us, as we are so used to using material objects to extend our own memories in terms of books, CDs, self-adhesive notes, and a whole host of other artefacts. But as soon as we try to imagine a world without such artificial memory devices, their role in our thinking, and especially creative thinking, becomes readily appreciated.

Quite what and how information was being stored in the material culture of the Upper Palaeolithic remains unclear. Some material culture may have simply acted to identify territorial boundaries, such as specific motifs that recur in several caves but have a restricted distribution in space (Sieveking 1979). The engraved designs on some artefacts, such as bone harpoons, may be marks of ownership or group affiliation (Conkey 1980). Those artefacts that appear to have a sequence of notches or incised lines may indeed be recording numbers of animals killed, people attending ceremonies, or the passing of seasons or years, as has been suggested for many years (reviewed in D'Errico and Cacho 1994). With regard to the animal imagery within cave paintings, I have argued that particular stylistic traits and subjects within the art function to cue the recall of knowledge about the natural world, and explicitly store some of that knowledge so that it can be transmitted to future generations (Mithen 1988, 1989, 1990). In these respects, the art of the Upper Palaeolithic is functioning to reduce the computational load on individual minds, expanding the possibilities of information storage, and enabling information and ideas to migrate between different individuals.

Material culture as a cognitive anchor

A second feature of the material culture of the Upper Palaeolithic, and indeed that of any period, is to allow new conceptual spaces that have a transient existence within a biological mind to be preserved, recreated within the biological mind, transferred into other minds, and then to be further transformed. To explain the role of material culture in this manner, we need to think once again about the evolutionary history of the mind.

One of the most puzzling features of modern minds is how they engage in domains of activity that have no adaptive value, and may even be maladaptive, whether in the modern world or the evolutionary environment of the mind. Two classic examples of this are religious thought and pure mathematics. While one may be able to identify particular functional values of particular elements of these, such as the creation of group cohesion resulting from the sharing of beliefs, quite how the seeming compulsion to create the conceptual spaces of supernatural beings or complex numbers arose remains unclear from an evolutionary perspective. My proposal is that such conceptual spaces emerged as a 'spandrel' (Gould and Lewontin 1979), as a non-functional by-product from an integration of the cognitive domains that had evolved in the early human mind (Mithen 1996b). Hence the concept of a supernatural being seems to be able to arise from taking ways of thinking about humans, animals and inert objects and integrating these in some fashion to create an idea of something that cannot exist in the real world.

Once such concepts are created, the problem is how to maintain them and to communicate them to other people. The communication of ideas about social relationships, tools or foraging would be relatively easy, as knowledge and ways of thinking about these domains have a deep evolutionary basis in the mind. We can appreciate this when we look at modern hunter-gatherers and see how such knowledge is transmitted so informally that one hardly realises it is happening (Mithen 1990: ch. 3). But religious ideas are far more difficult to share because they do not have a deep evolutionary basis in the mind. Consequently, these ideas need to become disembodied into durable media, and it is for this reason that material symbols are universally associated with religious ideas. The material symbols of these, whether we are dealing with the lion/man from Hohlenstein-Stadel, or Christ on a crucifix, serve to anchor the religious ideas – the conceptual spaces – into human minds. Without such anchors, the ideas would dissipate as they have no natural home within the mind (Mithen, 1997). Precisely the same argument can be made for the many other domains of thought that have no natural existence within the mind, such as pure mathematics.

The multivalency of material culture

The third critical feature of material culture, and one that further differentiates it from language to make it an even more powerful source of creative thinking, regards the fidelity of information transmission. When I receive an utterance from another person I do, of course, need to interpret that utterance to understand its meaning. If I am unclear what it means, and the receiver is still present, I can request clarification. Sometimes I will misunderstand but not be aware that I have misunderstood. This is, of course, a wonderful source for creative outputs, as is easily shown by the game of 'whispers'. (In this game, people are seated in a circle and the first person composes a message and whispers it to the one seated on his/her left. She/he then whispers the message to the next person, and so on until it goes completely around the circle. The message that returns to the one who composed it usually has no bearing on the original message at all.) Nevertheless, if the primary function of language is communication, perhaps originally about the social domain alone as discussed above, this implies a relatively high degree of fidelity with regard to the transmission of ideas and information.

Material culture has a far greater potential for ambiguity and the corruption of ideas. By definition, one cannot ask a rock face on which a painting is found, or a figurine carved out of ivory, for some clarification of the information and ideas they transmit. Consequently, it is most likely that the type of exploration and mapping of the conceptual spaces that the objects engender in each individual will be quite different. Many aspects of human behaviour can be interpreted as trying to avoid this – as attempts to maintain high fidelity in information transmission. The use of ritual is perhaps the most obvious example (Heyes 1993b). This often places people in highly emotional states or requires a rote repetition of ideas. As such, it ensures that essentially the same conceptual space is generated in different individuals and explored in the same manner.

On the other hand, the potential for the same item of material culture to generate multiple cognitive spaces is widely recognised and exploited in many societies. As has been described by Morphy (1989) and Faulstich (1992), for instance, images in the art of Australian Aborigines often have multiple meanings, and are interpreted differently by different individuals in different contexts. Consequently, a painting of a fish may facilitate one person to explore their cognitive spaces about fishing, while for another the conceptual space to be explored might be that about life and death, as fish can be potent symbols of rebirth. This multivalency most likely also applies to the art of the Upper Palaeolithic. Images of animals need not be *either* about hunting practices *or* about a mythological world, *or* a symbolic representation of human social relations: they were most probably about each of these at precisely the same time. This, of course, creates the basic dilemma we face as archaeologists when we search for the meanings of past paintings – there were probably as many meanings as there were people who viewed the art.

Material culture and the disembodied mind

We have seen, therefore, how material culture extends and disembodies the mind, enables ideas to migrate between individuals and vastly inflates the range of conceptual spaces that might exist, and the manner in which they can be explored and transformed. The remarkable efficacy of material culture in achieving this can be realised from the fact that we are only dimly aware that it is happening, other than at the times when we consult a dictionary or encyclopaedia.

The material culture of Early Humans had a limited role in this respect. It is likely that the maintenance of technical skill and the strong cultural traditions in tool morphology that we find among Early Humans arose partly from the fact that these were transmitted by the artefacts and passive observation of tool making alone, with no formal teaching by speech or gesture. This may indeed help to explain why they are often found in such large numbers at single sites – the idea to make tools was dependent upon tools being physically present. But due to the domain specific mentality of Early Humans, these artefacts could stimulate only a relatively fixed conceptual space to be explored, that about tool making. As soon as cognitive fluidity arose, the sight of a stone tool might facilitate the exploration of many new cognitive spaces.

I began this section by referring to material culture being as much a cause as a product of the behaviour and cognitive changes that underlie the Middle/Upper Palaeolithic transition. Now that we have seen how material culture disembodies the mind and facilities creative thought, the positive feedback loop with the behaviour and cognitive changes should be clear. Among Early Humans, children were born into a world devoid of material culture that cut across or combined different cognitive domains. This further supported the development of their domain-specific mentality during childhood. But as soon as objects existed in the world that cut across cognitive domains, such as images of supernatural beings, or of animals carved out of stone, this scaffolded the ontogenetic emergence of cognitive fluidity. Our lack of knowledge about childhood during the Palaeolithic, or more generally the evolution of ontogeny, is perhaps the most severe constraint on our understanding of cognitive evolution. But we can, I think, be confident that the material culture of the Upper Palaeolithic created trajectories of cognitive development very different from those of Early Humans, even if the initial brain substrate was essentially the same.

CONCLUSION

Modern humans are dramatically more creative in their behaviour than any other living species. When we examine the archaeological and fossil record, we find that they – we – are also dramatically more creative than any other of the types of humans that have lived since the emergence of the *Homo* genus 2.5 mya. Creative thinking does not appear to be simply a product of possessing a large brain, using speech, being reliant on material culture, occupying diverse types of environments or living in large social groups. Several types of humans – *H. erectus, H. neanderthalensis*, *archaic H. sapiens* – are characterised by such anatomy and behaviour but display what, with our modern hindsight, appears to be a remarkable monotony in their behaviour: the same types of stone tools repeatedly made over many thousands of years, the same foraging patterns, the same behavioural repertoires. We must, of course, appreciate that these were highly successful species and avoid using the absence of creative thought as the basis of a negative value judgement.

The first appearance of art 30,000 years ago is just one of a suite of new behaviours that suggest that a capacity for creative thinking, when defined in Boden's terms of exploring and transforming conceptual spaces, arose very late in human evolution, as much as 70,000 years after the first modern humans appeared. We can never be truly confident that we possess the earliest piece of art made by a human being, or indeed the trace of those other new behaviours. There are so many factors that led to the destruction of material culture, or the failure for its recovery, that to claim without doubt that art did not begin prior to 30,000 years ago is simply foolish. Buried in some cave or simply sitting on the surface of some exposed sediment there may indeed by a carved figurine or an engraved stone that, once found, would be so clearly a representation of an entity, or part of a symbolic code, that the art from Chauvet cave would look positively recent. But such artefacts have not as yet been found, if indeed they actually exist. Unfortunately, the arguments of those who have claimed that such artefacts have already been found – the incised bones from Bilzingsleben or the supposed figurine from Berekhet Ram – have been so deficient in rigour and methodology that the legitimate arguments as to why we should expect these artefacts to exist on purely theoretical grounds have been neglected. These arguments are, of course, the existence of anatomically modern humans at least 100,000 years ago (possibly much earlier in light of the evidence fom genetics), and modern brain size substantially earlier.

Yet even if art, or symbolic behaviour of some fashion, was found to date back to the first of the anatomically modern humans, leading us to conclude that creative thinking was generic to all members of our species as opposed to just those living after 30,000 years ago, its late appearance in human evolution would still be marked. There is unlikely to be one single change in the human mind that enabled conceptual spaces to become explored and transformed. Although creative thinking seems to appear suddenly in human evolution, its cognitive basis had a long evolutionary history during which the three foundations evolved largely on an independent basis: a theory of mind, a capacity for language and a complex material culture. After 50,000 years ago, these came to form the potent ingredients of a cognitive/social/material mix that did indeed lead to a creative explosion.

REFERENCES

Aiello, L. (1996a) 'Terrestriality, bipedalism and the origin of language.' In J. Maynard-Smith (ed.) *The Evolution of Social Behaviour Patterns in Primates and Man*, pp. 269–289. London: Proceedings of the British Academy 88.

Aiello, L. (1996b) 'Hominine preadaptations for language and cognition.' In P. Mellars and K. Gibson (eds) *Modelling the Early Human Mind*, pp. 89–99. Cambridge: McDonald Institute Monograph Series.

Aiello, L. and Dunbar, R.I.M. (1993) 'Neocortex size, group size and the evolution of language.' *Current Anthropology* 34: 184–193.
Allen, J. (1994) 'Radiocarbon determinations, luminescence dates and Australian archaeology.' *Antiquity* 68: 339–343.
Arensburg, B., Schepartz, L.A., Tillier, A.M., Vandermeersch, B. and Rak, Y. (1990) 'A reappraisal of the anatomical basis for speech in Middle Palaeolithic hominids.' *American Journal of Physical Anthropology* 83: 137–146.
Bahn, P. and Vertut, J. (1988) *Images of the Ice Age*. London: Windward.
Bar-Yosef, O. (1994) 'The contributions of southwest Asia to the study of the origin of modern humans.' In M.H. Nitecki and D.V. Nitecki (eds) *Origins of Anatomically Modern Humans*, pp. 23–66. New York: Plenum Press.
Baron-Cohen, S. (1995) *Mindblindness*. Cambridge, MA: MIT Press.
Baron-Cohen, S. (in press) 'The evolution of a theory of mind' In M. Corballis (ed.) *The Evolution of the Hominid Mind*. Oxford: Oxford University Press.
Bednarik, R.G. (1992) 'Palaeoart and archaeological myths.' *Cambridge Archaeological Journal* 2: 27–57.
Bednarik, R.G. (1994) 'A taphonomy of palaeoart.' *Antiquity* 68: 68–74.
Bednarik, R.G. (1995) 'Concept mediated marking in the Lower Palaeolithic.' *Current Anthropology* 36: 605–634.
Begun, D. and Walker, D. (1993) 'The endocast.' In D. Walker and R. Leakey (eds) *The Nariokotome* Homo Erectus *Skeleton*, pp. 326–358. Berlin: Springer Verlag.
Bickerton, D. (1990) *Language and Species*. Chicago: Chicago University Press.
Bickerton, D. (1996) *Language and Human Behaviour*. London: UCL Press.
Bloom, P. (in press) 'Some issues in the evolution of language and thought.' In D. Cummins and C. Allen (eds) *Evolution of the Mind*. Cambridge, MA: MIT Press.
Boden, M. (1990) *The Creative Mind: Myths and Mechanisms*. London: Weidenfeld & Nicolson.
Boden, M. (1994) 'Précis of *The Creative Mind: Myths and Mechanisms*.' *Behavioural and Brain Sciences* 17: 519–570.
Boesch, C. and Boesch, H. (1989) 'Hunting behaviour of wild chimpanzees in the Taï National Park.' *American Journal of Physical Anthropology*, 78: 547–573.
Boyer, P. (1994) *The Naturalness of Religious Ideas*. Berkeley: University of California Press.
Breuil, H. (1952) *Four Hundred Centuries of Cave Art*. Montignac: Centre d'Etudes et de Documentation Préhistoriques.
Byrne, R.W. (1995) *The Thinking Ape: Evolutionary Origins of Intelligence*. Oxford: Oxford University Press.
Byrne, R.W. and Whiten, A. (eds) (1988) *Machiavellian Intelligence: Social Expertise and the Evolution of Intellect in Monkeys, Apes and Humans*. Oxford: Clarendon Press.
Carruthers, P. (in press) 'Thinking in language? Evolution and a modularist possibility.' In P. Carruthers and J. Boucher (eds) *Language and Thought: Interdisciplinary Themes*. Cambridge: Cambridge University Press.
Chase, P. (1986) *The Hunters of Combe Grenal: Approaches to Middle Palaeolithic Subsistence in Europe*. Oxford: British Archaeological Reports, International Series, 286.

Chase, P. and Dibble, H. (1987) 'Middle Palaeolithic symbolism: a review of current evidence and interpretations.' *Journal of Anthropological Archaeology* 6: 263–293.

Chase, P. and Dibble, H. (1992) 'Scientific archaeology and the origins of symbolism: a reply to Bednarik.' *Cambridge Archaeological Journal* 2: 43–51.

Chauvet, J.-M., Deschamps, E.B. and Hillaire, C. (1996) *Chauvet Cave: The Discovery of the World's Oldest Paintings*. London: Thames & Hudson.

Clark, A. (1996) *Being There: Putting Brain, Body and World Together Again*. Cambridge, MA: MIT Press.

Clottes, J. (1990) 'The parietal art of the Late Magdalenian.' *Antiquity* 64: 527–548.

Clottes, J. (1996) 'Thematic changes in Upper Palaeolithic art: a view from the Grotte Chauvet.' *Antiquity* 70: 276–288.

Conkey, M. (1980) 'The identification of prehistoric hunter-gatherer aggregation sites.' *Current Anthropology* 21: 609–630.

Corballis, M.C. (1992) 'On the evolution of language and generativity.' *Cognition* 44: 197–226.

Davidson, I. (1991) 'The archaeology of language origins: A review.' *Antiquity* 65: 39–48.

Davidson, I. (1992) 'There's no art – To find the mind's construction – In offence (reply to R. Bednarik).' *Cambridge Archaeological Journal* 2: 52–57.

Davidson, I. and Noble, W. (1989) 'The archaeology of perception: traces of depiction and language.' *Current Anthropology* 30: 125–155.

Davidson, I. and Noble, W. (1992) 'Why the first colonisation of the Australian region is the earliest evidence of modern human behaviour.' *Archaeology in Oceania* 27: 113–119.

Dennett, D. (1988) 'The intentional stance in theory and practice.' In R.W. Byrne and A. Whiten (eds) *Machiavellian Intelligence: Social Expertise and the Evolution of Intellect in Monkeys, Apes and Humans*, pp. 180–202. Oxford: Clarendon Press.

D'Errico, F. (1991) 'Microscopic and statistical criteria for the identification of prehistoric systems of notation.' *Rock Art Research* 8: 83–93.

D'Errico, F. (1995) 'A new model and its implications for the origin of writing: the la Marche antler revisited.' *Cambridge Archaeological Journal* 5: 163–206.

D'Errico, F. and Cacho, C. (1994) 'Notation versus decoration in the Upper Palaeolithic; A case study from Tossal de la Roca, Alicante, Spain.' *Journal of Archaeological Science* 21: 185–200.

Donald, M. (1991) *Origins of the Modern Mind*. Cambridge, MA: Harvard University Press.

Dunbar, R.I.M. (1993) 'Coevolution of neocortical size, group size and language in humans.' *Behavioural and Brain Sciences* 16: 681–735.

Dunbar, R.I.M. (1996) *Grooming, Gossip and the Evolution of Language*. London: Faber & Faber.

Falk, D. (1983) 'Cerebral cortices of East African early hominids.' *Science* 221: 1072–1074.

Faulstich, P. (1992) 'Of earth and dreaming: abstraction and naturalism in Warlpiri art.' In M. J. Morwood and D.R. Hobbs (eds) *Rock Art and Ethnography*, pp. 19–23. Melbourne: Occasional AURA Publication No. 5.

Fullagar, R.L.K., Price, D.M. and Head, L.M. (1996) 'Early human occupation of northern Australia: Jinmium rock-shelter.' *Antiquity* 70: 751–773.

Gamble, C. (1993) *Timewalkers: The Prehistory of Global Colonization*. Stroud: Alan Sutton.

Girard, C. and David, F. (1982) 'A propos de la chase spécialisée au Paléolithique moyen: l'exemple de Mauran (Haute-Garonne).' *Bulletin de la Société Préhistorique Française* 79: 11–12.

Goren-Inbar, N. (1986) 'A figurine from the Acheulean site of Berekhat Ram.' *Mi'tekufat Ha'even* 19: 7–12.

Gould, S.J. and Lewontin, R.C. (1979) 'The spandrels of San Marco and the panglossian paradigm: a critique of the adaptionist programme.' *Proceedings of the Royal Society of London* B 205: 581–598.

Hahn, J. (1993) 'Aurignacian art in Central Europe.' In H. Knecht, A. Pike-Tay and R. White (eds) *Before Lascaux: The Complex Record of the Early Upper Palaeolithic*, pp. 229–241. Boca Ralton: CRC Press.

Heyes, C.M. (1993a) 'Anecdotes, training and triangulating: do animals attribute mental states?' *Animal Behaviour* 46: 177–188.

Heyes, C.M. (1993b) 'Imitation, culture and cognition.' *Animal Behaviour* 46: 999–1010.

Holloway, R.L. (1981a) 'Culture, symbols and human brain evolution.' *Dialectical Anthropology* 5: 287–303.

Holloway, R.L. (1981b) 'Volumetric and asymmetry determinations on recent hominid endocasts: Spy I and II, Djebel Irhoud I, and Sale *Homo erectus* specimens, with some notes on Neanderthal brain size.' *American Journal of Physical Anthropology* 55: 385–393.

Holloway, R.L. (1985) 'The poor brain of *Homo sapiens neanderthalensis*: See what you please . . .' In E. Delson (ed.)*Ancestors: The Hard Evidence*, pp. 319–324. New York: Alan R. Liss.

Houghton, P. (1993) 'Neanderthal supralaryngeal vocal tract.' *American Journal of Physical Anthropology* 90: 139–146.

Humphrey, N. (1976) 'The social function of intellect.' In P.P.G. Bateson and R.A. Hinde (eds) *Growing Points in Ethology*, pp. 303–317. Cambridge: Cambridge University Press.

Klein, R.G. (1989) 'Biological and behavioural perspectives on modern human origins in Southern Africa.' In P. Mellars and C. Stringer (eds) *The Human Revolution*, pp. 530–546. Edinburgh: Edinburgh University Press.

Koestler, A. (1964) *The Act of Creation*. New York: Macmillan.

Leroi-Gourhan, A. (1965) *Préhistoire de l'Art Occidental*. Paris: Mazenod.

Leroi-Gourhan, A. (1983) 'Les entités imaginaires. Esquisse d'une recherche sur les monstres pariétaux paléolithiques.' In *Homenage al Prof. M. Almargo Basch, Vol. 1*, pp. 251–263. Madrid: Ministerio de Cultura.

Leslie, A. (1991) 'The theory of mind impairment in autism: evidence for a modular mechanism of development.' In A. Whiten (ed.) *Natural Theories of Mind: Evolution, Development and Simulation of Everyday Mindreading*, pp. 63–78. Oxford: Basil Blackwell.

Lorblanchet, M. and Sieveking, A. (1997) 'The monsters of Pergouset.' *Cambridge Archaeological Journal* 7: 37–56.

Mania, D. and Mania, U. (1988) 'Deliberate engravings on bone artefacts by *Homo erectus*.' *Rock Art Research* 5: 91–107.

Marshack, A. (1972) *The Roots of Civilization*. London: McGraw Hill.

Marshack, A. (1990) 'Early hominid symbols and the evolution of human capacity.' In P. Mellars (ed.) *The Emergence of Modern Humans*, pp. 457–498. Edinburgh: Edinburgh University Press.

Marshack, A. (1991) 'The Täi plaque and calendrical notation in the Upper Palaeolithic.' *Cambridge Archaeological Journal* 1: 25–61.

Marshack, A. (1997) 'The Berekhat Ram figurine: a late Acheulian carving from the Middle East.' *Antiquity* 71: 327–337.

Mellars, P. (1973) 'The character of the Middle–Upper transition in south-west France.' In C. Renfrew (ed.) *The Explanation of Culture Change*, pp. 255–276. London: Duckworth.

Mellars, P. (1989a) 'Technological changes at the Middle–Upper Palaeolithic transition: economic, social and cognitive perspectives.' In P. Mellars and C. Stringer (eds) *The Human Revolution*, pp. 338–365. Edinburgh: Edinburgh University Press.

Mellars, P. (1989b) 'Major issues in the emergence of modern humans.' *Current Anthropology* 30: 349–385.

Mellars, P. (1991) 'Cognitive changes and the emergence of modern humans.' *Cambridge Archaeological Journal* 1: 63–76.

Mithen, S. (1988) 'Looking and learning: Upper Palaeolithic art and information gathering.' *World Archaeology* 19: 297–327.

Mithen, S. (1989) 'To hunt or to paint? Animals and art in the Upper Palaeolithic.' *Man* 23: 671–695.

Mithen, S. (1990) *Thoughtful Foragers: A Study of Prehistoric Decision Making*. Cambridge: Cambridge University Press.

Mithen, S. (1993) 'From domain-specific to generalised intelligence: a cognitive interpretation of the Middle/Upper Palaeolithic transition.' In C. Renfrew and E. Zubrow (eds) *The Ancient Mind*, pp. 29–39. Cambridge: Cambridge University Press.

Mithen, S. (1996a) 'On Early Palaeolithic "concept-mediated marks", mental modularity and the origins of art.' *Current Anthropology* 37: 666–670.

Mithen, S. (1996b) *The Prehistory of the Mind: A Search for the Origins of Art, Science and Religion*. London and New York: Thames & Hudson.

Mithen, S. (1996c) 'Domain-specific intelligence and the Neanderthal mind'. In P. Mellars and K. Gibson (eds) *Modelling the Early Human Mind*, pp. 217–229. Cambridge: McDonald Institute Monographs.

Mithen, S. (1997) 'Cognitive archaeology, evolutionary psychology and cultural transmission with particular reference to religious ideas.' In C. Michael Barton and G.A. Clark (eds) *Rediscovering Darwin: Evolutionary Theory in Archaeological Explanation*, pp. 67–74. *Archaeological Papers of the American Anthropological Association No. 7.*

Morphy, H. (1989) 'On representing Ancestral Beings.' In H. Morphy (ed.) *Animals into Art*, pp. 144–160. London: Unwin Hyman.

Pelcin, A. (1994) 'A geological explanation for the Berekhat Ram figurine.' *Current Anthropology* 35: 674–675.

Perkins, D. (1994) 'Creativity: beyond the Darwinian paradigm.' In M. Boden (ed.) *Dimensions of Creativity*, pp. 119–142. Cambridge MA: MIT Press.

Pfeiffer, J. (1982) *The Creative Explosion*. New York: Harper & Row.

Pigeot, N. (1990) 'Technical and social actors: flint knapping specialists and apprentices at Magdalenian Etiolles.' *Archaeological Review from Cambridge* 9: 126–141.

Roberts, M.B. (1986) 'Excavation of the Lower Palaeolithic site at Amey's Eartham Pit, Boxgrove, West Sussex: a preliminary report.' *Proceedings of the Prehistoric Society* 52: 215–246.

Ruff, C.B., Trinkaus, E. and Holliday, T.W. (1997) 'Body mass and ecaphalization in Pleistocene *Homo*.' *Nature* 387: 173–176.

Schepartz, L.A. (1993) 'Language and modern human origins.' *Yearbook of Physical Anthropology* 36: 91–126.

Scott, F.J. and Baron-Cohen, S. (1996) 'Imagining real and unreal things: evidence of a dissociation in autism.' *Journal of Cognitive Neuroscience* 8: 371–382.

Scott, K. (1980) 'Two hunting episodes of Middle Palaeolithic age at La Cotte de Saint-Brelade, Jersey (Channel Islands).' *World Archaeology* 12: 137–152.

Sieveking, A. (1979) 'Style and regional groupings in Magdalenian cave art.' *Bulletin of the Institute of Archaeology* 16: 95–109.

Soffer, O. (1994) 'Ancestral lifeways in Eurasia – The Middle and Upper Palaeolithic records.' In M.H. Nitecki and D.V. Nitecki (eds) *Origins of Anatomically Modern Humans*, pp. 101–120. New York: Plenum Press.

Stringer, C. and Gamble, C. (1993) *In Search of the Neanderthals*. London: Thames & Hudson.

Thieme, H. (1997) 'Lower Palaeolithic hunting spears from Germany.' *Nature* 385: 807–810.

Tobias, P.V. (1987) 'The brain of *Homo habilis*: a new level of organisation in cerebral evolution.' *Journal of Human Evolution* 16: 741–761.

Valladas, H., Cachier, H., Maurice, P., Bernaldo de Quiros, F., Clottes, J., Caberera Valdés, V., Uzquiano, P. and Arnold, M. (1992) 'Direct radiocarbon dates for prehistoric paintings at the Altamira, El Castillo and Niaux caves.' *Nature* 357: 68–70.

Weisberg, R.W. (1993) *Creativity: Beyond the Myth of Genius*. New York: W.H. Freeman.

Whallon, R. (1989) 'Elements of culture change in the Later Palaeolithic.' In P. Mellars and C. Stringer (eds) *The Human Revolution*, pp. 433–454. Edinburgh: Edinburgh University Press.

White, R. (1982) 'Rethinking the Middle/Upper Palaeolithic transition.' *Current Anthropology* 23: 169–192.

White, R. (1989) 'Production complexity and standardization in early Aurignacian bead and pendant manufacture: evolutionary implications.' In P. Mellars and C. Stringer (eds) *The Human Revolution*, pp. 366–390. Edinburgh: Edinburgh University Press.

Whiten, A. (ed.) (1991) *Natural Theories of Mind: Evolution, Development and Simulation of Everyday Mindreading*. Oxford: Basil Blackwell.

Wynn, T. (1991) 'Tools, grammar and the archaeology of cognition.' *Cambridge Archaeological Journal* 1: 191–206.

PART III

Creativity in later prehistoric Europe

CHAPTER ELEVEN

INTRODUCTION TO PART III

The four chapters in this part examine how the evidence for later prehistoric periods in Europe can help us to understand the phenomenon of creativity. In doing this, they also explore whether, and if so, how, we can gain greater insights into past cultures and culture change by structuring the questions we ask about the past around the notion of creativity and by adopting an explicitly cognitive approach.

By 'later prehistory' I am referring to the period after the end of the last Ice Age, conventionally placed at 10,000 years ago, and prior to the expansion of the Roman Empire across Europe, a little over 2,000 years ago. Just as in the previous section, with only four chapters a mere fraction of the data and issues raised by this long and complex period can be addressed. Nevertheless, the chapters tackle some of the most important cultural developments: the first appearance of formal cemeteries, the construction of megalithic tombs, the production of monumental temples and figurative art, and changes in man–animal relationships with regard to the use of the horse in peace and war.

As with the chapters dealing with our evolutionary history, these contributions need to be placed into a chronological context if they are fully to be appreciated. By so doing, additional issues of later European prehistory that are ripe for a similar treatment of an explicitly cognitive approach, with the notion of human creativity at its core, can be readily identified. Fortunately, there have been several excellent syntheses of European later prehistory, and specific periods therein, to which those readers not familiar with how the subjects of this section fit into the larger picture of European prehistory can turn. Particularly recommended is *The Oxford Illustrated Prehistory of Europe*, edited by Barry Cunliffe (1994). Other useful books to place these chapters into their context are referred to below.

To start 'later' European prehistory at the end of the last Ice Age is quite arbitrary. Although this was a period of dramatic environmental change, there is significant cultural continuity across the Pleistocene/Holocene boundary, reflecting the ability of prehistoric hunter-gatherers to adapt to radically changing environments. The means of that adaptation was human decision making in the face of environmental uncertainty (cf. Mithen 1990). It is perhaps here, right at the start of later prehistory, that the human capacity for creative thinking was pushed to its limits, as it is becoming increasingly apparent that some of the environmental changes were catastrophic in nature (e.g. Blanchon and Shaw 1995; Dansgaard *et al.* 1989).

Examining this period of human adaptation from a more cognitive perspective is a task for the future, as this volume moves straight to those hunter-gatherers who had become established in the new, thickly forested, post-glacial environments of Europe – hunter-gatherers of the Mesolithic (Bonsall 1989; Mithen 1994; Zvelebil 1986). This had once been seen as a 'dark age', culturally eclipsed by the earlier achievements of the Upper Palaeolithic cave painters and those of the Neolithic crop planters that were to come (Rowley-Conwy 1986). We now appreciate, however, that the Mesolithic hunter-gatherers of Europe were just as culturally creative as those societies that came before and after. In the few thousand years that they existed in Europe, a remarkable suite of technological innovations occurred including pottery, fish traps, sophisticated bows, and plant-processing technology. They had a spectacular art, but much of it appears to have been made on organic materials, constraining the amount that has survived in the archaeological record. Mesolithic hunter-gatherers were also innovative in social behaviour. By the later Mesolithic period, it seems likely that those groups within the resource-rich ecozones of coasts and large rivers were living a lifestyle not normally associated with hunting and gathering: they were largely sedentary and egalitarianism was fast disappearing (Rowley-Conwy 1983; Price 1985).

One of the most dramatic types of evidence for these social developments is the appearance of formal burial cemeteries in many parts of Europe after 7,000 years ago. It is these cemeteries, and the creativity that may be evident in the various acts of burial, that Rick Schulting addresses in his chapter. As he describes, the variability in the practices relating to death during the later Mesolithic is almost bewildering. He suggests that we should view these as a series of creative responses to the threatened disruption of society caused by the death of its members. Burial ritual provides a context either to reinforce, or to challenge, the existing sociopolitical order.

Schulting's chapter shows how far Mesolithic studies have moved in the last decade; no longer is the period just about coping with environmental change – it is seen as providing a key data set for understanding social and ideological behaviour in human communities. The emergence of complex hunter-gatherers (cf. Price and Brown 1985) during the Early Holocene is perhaps the first of the major developments within later European prehistory. The second is the spread and indigenous development of farming leading to the eventual demise of hunters and gatherers in Europe.

The establishment of farming communities across Europe, traditionally associated with the cultural package of monument construction, pottery and ground stone axes that constitute the Early Neolithic, is a complicated story. It involved the spread of new people, the spread of new ideas, and innovations by the indigenous Mesolithic communities. The relationship between agricultural lifestyles and the Neolithic is not as clear cut as once had been thought, as changes in material culture and subsistence economy were not necessarily related. The archaeological record shows considerable variability from region to region in terms of timing, process and the eventual economy that emerged. Alasdair Whittle (1996) has provided an excellent synthesis of Europe during the Neolithic, while the spread of farming and its consequences have been explored from different perspectives by Renfrew (1987) and Hodder (1990).

For Europe as a whole, it took over 3,000 years for farming to become established after the first farmers appeared in southeast Europe 9,000 years ago. In some areas, the first farmers were sedentary, with cereals crops and cattle, pottery and ground stone axes – the classic 'Neolithic' package that had once been thought to characterise early farmers across the whole of Europe. In other areas, the Early Neolithic communities possessed a lifestyle that had more in common with the hunter-gatherers of the Mesolithic than with our traditional ideas of farming. Indeed, they may have been substantially more mobile than those Late Mesolithic communities which Schulting discussed.

One area of Europe where this is the case is on the Atlantic fringe. At this far western edge of the Eurasian landmass, the start of the Neolithic and of farming is associated with the construction of megalithic tombs. These are found in Portugal, France, Britain and south Scandinavia, and seem to reflect a new way of structuring relationships between people, and between people and the land. Richard Bradley has written extensively about such megaliths, and in the second contribution of this part he considers one particular group of these, the Clava Tombs from Scotland, where he undertook field work between 1994–97.

In his chapter, Bradley begins by immediately invoking one of the major themes discussed by Boden: that often, perhaps always, creative thinking requires the discipline of rules. Imagination without constraint is unlikely to produce novel items of value that are socially acceptable. Bradley explains how the Clava Tombs can be understood as arising from the 'interplay between cosmology and engineering' – with the implication that this is of general significance for understanding megalithic architecture, and indeed these Early Neolithic societies as a whole. If the cosmological world was the conceptual space being explored, then the constraints arose from what could be done with unshaped lumps of stone as the building material and medium of expression. So megaliths – such a striking creative achievement when seen in the Scottish landscapes at Clava – arose from the tension between combining abstract ideas about the cosmos with the earthy reality of erecting monuments that have no choice but to obey the laws of physics.

By 4,000 years ago, considerable cultural and economic variability had emerged throughout Europe. Sedentary mixed crop agriculturists occupied long houses on the European plain, lakeside dwellings had been built in Switzerland, and tells developed in southeast Europe (Whittle 1996). Within and between these communities, trade and exchange systems were active, involving such items as the shells of *Spondylus gaederoups*, originating in the east Mediterranean, and axes made from jade. One of the underlying economic developments was the use of 'secondary products' – milk, wool and traction from domesticated animals – as has been explored by Andrew Sherrat (1997) in a series of stimulating essays. Exploiting these products, making objects for trade, developing exchange networks and constructing new types of dwellings all involved creative thinking; so too did new technological innovations such as the copper smelting that arose independently in the Balkans and in Iberia.

One theme of this later Neolithic period of European prehistory is increasing cultural variability – different regions can be seen to undergo their own particular cultural trajectories due to their own social, historical, economic and environmental conditions. One of the most startling independent developments is examined by Caroline Malone and Simon Stoddart in Chapter 14: the remarkable case of Malta. This tiny speck of an island in the Mediterranean developed a culture during the Late Neolithic quite unlike anywhere else and provides an excellent case study in the prehistory of creative thought and action.

As with Bradley, Malone and Stoddart draw on their own field work within their chapter. And, like other contributors, they emphasise that an understanding of human creativity requires that the actions of individuals and groups are firmly embedded into a social context. Their focus is the period on Malta between 6,000 and 4,000 years ago, during which spectacular temples were constructed and a remarkable range of figurative art objects produced. They describe how during the last century this creative achievement has been rather underplayed by archaeologists, who have tried to find the source of these developments outside of Malta, such as in North Africa, or imposed inappropriate models developed for later periods, or indeed spurious ideas about 'mother goddess' cults. As such, they illustrate the argument I made in the introduction to this book – that archaeologists have too often shied away from dealing with human creativity.

Stoddart and Malone explain how the key factor in understanding the cultural developments on Malta was its isolation from societies elsewhere in the Mediterranean. As with the Mesolithic burials and construction of megalithic tombs, the ultimate motivating factor was religion. This religious fervour was channelled through social rivalry between families who initially competed for exotic raw materials, and then in the construction of temple structures. The role of constraint is again discussed in the sense that, within public areas, sets of rules limited the variability in

Maltese art. But in the more private areas there were more opportunities for individual expression.

As European culture developed, moving on from the appearance and consolidation of farming into the Bronze Age, the extent of cultural diversity increased. Problems of environmental change, some of anthropogenic origin such as soil erosion, and some induced by climatic change, problems of population growth and of social stress, led to economic intensification and social evolution. In Mediterranean regions, the vine and olive were incorporated into the economy, and in some places irrigation systems established. Such developments were partly responsible for settlement nucleation and fortification, and the development of hierarchical societies. The ultimate consequence was the emergence of the first European civilisations of Minoan Crete and Mycenaean Greece. Excellent studies of this period in southern Europe are provide by Dickinson (1994) for the Aegean, Chapman (1990) for Iberia, and in the edited volume by Mathers and Stoddart (1994) for the Mediterranean as a whole. In other regions, agricultural intensification took on more varied forms, and a set of elites and chiefdoms was established, competing for the supply of various prestigious items that acted as symbols of power. Kristiansen (1987) and Shennan (1993) provide useful review essays for such developments in Scandinavia and central Europe respectively.

Soon after 3,000 years ago, society began to undergo a major transformation that was both a cause and a consequence of the dramatic developments in metallurgical production. This ushered in the Late Bronze Age period, to be followed by the Iron Age, which continued until the expansion of the Roman Empire. Syntheses of these periods are provided by Andrew Sherrat, Anthony Harding, Barry Cunliffe and Timothy Taylor, within *The Oxford Illustrated Prehistory of Europe* (Cunliffe 1994), while Collis (1984) has produced an important review of the Iron Age in Europe.

The new metallurgy was producing a vast array of objects, ranging from figurines to swords, on an unprecedented scale. Initially these were in copper, then in bronze, and after 3,000 years ago were principally in iron. Agricultural systems became more intensified, partly reflected in the imposition of greater numbers of land boundaries and field systems, especially in the Atlantic regions of Europe. Throughout Europe there was a general transformation in burial practices as cremations replaced inhumation. Trade and exchange systems continued to expand, and although the direct evidence for ships is limited, there can be little doubt that long sea voyages for trading were made. Many cultural developments seem to indicate that warfare was becoming both more common and larger in scale. Indeed, many of the bronze and iron items were either weapons or armour, while settlements throughout Europe became increasingly fortified.

This brings us to the final contribution to this volume, a discussion of the changing role of the horse during these later prehistoric periods by Colin Renfrew. Some of its roles were clearly military – pulling chariots and supporting cavalrymen– and Renfrew describes his chapter as examining creative innovations in the field of warfare. But he uses this specific topic to support a more general argument, that to understand change in prehistoric societies one needs to understand the 'cognitive constellations' of both individuals and groups. To use the horse in a military role, for instance, requires for that role to be part of the mindset of the people – one cannot simply assume this is the case as soon as domesticated horses are present in the archaeological record. There has been an assumption, he argues, that mounted warrior horsemen were present from the start of the Bronze Age; but by examining the iconography of the period, Renfrew knows that this was not the case.

Renfrew's chapter follows others in this volume by stressing the need to understand creativity, and more particularly innovation, in its social context. His notion of a cognitive constellation as a habitual association of ideas is one that could be applied to other case studies. Schulting, for instance, identifies recurring associations between mandibles and triangular settings of stone in the Iron Gates region of Mesolithic Europe. The manner in which Renfrew describes how an individual's 'mappa' is partly structured by the experience of the manmade world, by material culture, echoes the argument in my contribution within part II that when dealing with modern humans we cannot separate mind and material culture in a meaningful way.

Renfrew's chapter traces the changing roles of the horse to the end of the prehistoric period in Europe. By the Late Iron Age, temperate Europe had become transformed from the nucleated farmsteads of the first Neolithic farmers to settlements most effectively described as towns, and social formations described as states (Collis 1984). Centres of industrialised production existed and coinage had been introduced. But in the Mediterranean, developments had been even more dramatic, with the emergence of the city-states with powerful military organisations. The Athenian state and then empire expanded in the fifth century AD, followed by that of Macedonia. But by the fourth century, the most powerful of all was emerging, Rome, which would soon extend its empire and bring European prehistory to a close.

Later European prehistory is a remarkable source of data and ideas for those wishing to study the prehistory of human creativity. The chapters in this section provide just four case studies, out of a vast number that need to be undertaken. For if we are to understand the adaptations to the post-glacial climates, the adoption of domesticated resources, the technological innovations ranging from pottery to iron working, the emergence of elites and state society, we must, at some time, grasp the nettle, and examine the role and nature of human creativity. We must, however, consider this not as a flash of genius nor as an isolated mental phenomenon excluding study of environment, economy and society. For creativity is the stuff of day-today life: just as it cannot be divorced from mental processes, neither can it be separated from bodily actions and needs, from social and economic behaviour, nor from that which is the essence of archaeological study – material culture.

REFERENCES

Blanchon, P. and Shaw, J. (1995) 'Reef drowning during the last deglaciation: evidence for catastrophic sea-level rise and ice sheet collapse.' *Geology* 23: 4–8.

Bonsall, C. (ed.) (1989) *The Mesolithic in Europe*. Edinbugh: John Donald.

Chapman, P.W. (1990) *Emerging Complexity: The Later Prehistory of South-East Spain, Iberia and the West Mediterranean*. Cambridge: Cambridge University Press.

Collis, J. (1984) *The European Iron Age*. London: Batsford.

Cunliffe, B. (ed.) (1994) *The Oxford Illustrated Prehistory of Europe*. Oxford: Oxford University Press.

Dansgaard, W., White, J.W.C. and Johnsen, S.J. (1989) 'The abrupt termination of the Younger Dryas climate event.' *Nature* 339: 532–534.

Dickinson, O. (1994) *The Aegean Bronze Age*. Cambridge: Cambridge University Press.

Hodder, I. (1990) *The Domestication of Europe*. Oxford: Blackwell.

Kristiansen, K. (1987) 'Centre and periphery in Bronze Age Scandinavia.' In M. Rowlands, M.T. Larsen and K. Kristiansen (eds) *Centre and Periphery in the Ancient World*, pp. 74–85. Cambridge: Cambridge University Press.

Mathers, C. and Stoddart, S. (eds) (1994) *Development and Decline in the Mediterranean Bronze Age*. Sheffield Archaeological Monographs no. 8. Sheffield: J.R. Collins Publications.

Mithen, S. (1990) *Thoughtful Foragers: A Study of Prehistoric Decision Making*. Cambridge: Cambridge University Press.

Mithen, S. (1994) 'The Mesolithic age.' In B. Cunliffe (ed.) *The Oxford Illustrated Prehistory of Europe*, pp. 79–135. Oxford: Oxford University Press.

Price, T.D. (1985) 'Affluent hunters of Mesolithic southern Scandinavia.' In T.D. Price and J.A. Brown (eds) *Prehistoric Hunter-Gatherers: The Emergence of Cultural Complexity*, pp. 341–362. New York: Academic Press.

Price, T.D. and Brown, J.A. (eds) (1985) *Prehistoric Hunter-Gatherers: The Emergence of Cultural Complexity*. New York: Academic Press.

Renfrew, C. (1987) *Archaeology and Language: The Puzzle of Indo-European Origins*. London: Jonathan Cape.

Rowley-Conwy, P. (1983) 'Sedentary hunters: the Ertebølle example.' In G.N. Bailey (ed.) *Hunter-Gatherer Economy in Prehistory*, pp. 111–126. Cambridge: Cambridge University Press.

Rowley-Conwy, P. (1986) 'Between crop planters and cave painters: aspects of the temperate European Mesolithic.' In M. Zvelebil (ed.) *Hunters in Transition*, pp. 17–32. Cambridge: Cambridge University Press.

Shennan, S. (1993) 'Commodities, transactions and growth in the Central European Early Bronze Age.' *Journal of European Archaeology* 1(2): 59–72.

Sherrat, A. (1997) *Economy and Society in Prehistoric Europe*. Edinburgh: Edinburgh University Press.

Whittle, A. (1996) *Europe in the Neolithic*. Cambridge: Cambridge University Press.

Zvelebil, M. (ed.) (1986) *Hunters in Transition*. Cambridge: Cambridge University Press.

CHAPTER TWELVE

CREATIVITY'S COFFIN

Innovation in the burial record of Mesolithic Europe

RICK J. SCHULTING

INTRODUCTION: DEATH AND CREATIVITY

Creativity can be thought of as a latent quality or capacity that becomes expressed under certain conditions. It is the ability to combine existing forms in new ways, finding novel solutions to problems. Yet it is more than this as well – it is, as Margaret Boden (1990) refers to it, the opening up of new conceptual spaces, in which new thoughts become possible. The treatment of the dead may seem a strange place to look for 'creativity' – funerary behaviour is often cited as being among the more conservative aspects of a society. Yet the burial record of Upper Palaeolithic and, more particularly, Mesolithic Europe exhibits remarkable diversity, including inhumation, cremation, boat burial, 'skull nests', the use of complex grave structures of stone, antler and wood, and the circulation of skeletal parts. In southern Scandinavia, dogs were accorded elaborate burial treatment equivalent to, or in some cases surpassing, that given to humans. The variety of practices relating to death and burial can be viewed as a series of creative responses to the threatened disruption of society by the death of its members, and, more particularly, to the opportunities provided to reinforce or challenge the existing sociopolitical order. This is the view of mortuary behaviour that I will concentrate on here. It is not the only view – funerary rites have other emotive and social uses and meanings (Hertz 1907, van Gennep 1960) – but I feel that it is one that is more susceptible to interpretation in terms of creative response to perceived opportunity.

While it has been recognised that 'natural' symbols do not exist, the human body in many cultures forms a strong metaphorical basis for cosmological, social and political constructions (and vice versa) (Douglas 1996). The corpse can also be a powerful source of symbols (Metcalf and Huntington 1991). Harnessing this potential provides a greater potency to relationships and negotiations worked out between the living and the dead, as well as between various individuals and groups within the living community. As an example one can look at the eagerness with which the body parts of saints – relics – were sought in the medieval period. They were stolen, sold, traded and created, and they were embedded in political relationships between the Church and its sometimes wayward allies (Geary 1986). Acts involving the body, whether living or dead, possess a great capacity to engage us.

The central focus of this chapter is the social context of creativity: an issue not adequately covered in Boden's (1990) work but one of considerable significance (cf. Simonton 1994 and see Mithen, Malone and Stoddart, and Renfrew, this volume). It is in the study of the social context in which creativity occurs that archaeology can make a significant contribution. One of the most striking features of the archaeological record is that the rate of change in different times and places is extraordinarily variable. There are many examples of long periods of stability exhibiting only gradual cumulative changes, punctuated by brief periods exhibiting rapid change and innovation. The potential for creativity is constantly present, but its expression is variable; under certain social conditions, innovations will be more positively valued, and ideas that would have been ignored under other circumstances are now encouraged and followed through. This in turn provides incentive for further creative expression (Renfrew [1978] has expressed similar ideas regarding the context of the *spread* of new ideas and inventions). I will argue in this chapter that a relevant context for creativity in burial practices is one of developing social and economic status, rivalry and competition.

Creativity as it is studied today emphasises the individual mind (Boden 1990). This puts archaeologists at somewhat of a disadvantage, since the actions of specific individuals, though we know they are represented in the materials we recover, are very hard to distinguish. Is it possible to identify the 'first' time a particular behaviour appeared? Attempts at finding firsts have been popular but notoriously unsuccessful in archaeology. This caveat should be kept in mind when I speak of the 'first' appearance of such things as boat burials or cremations. There is no way of knowing how innovative these behaviours actually were – indeed, they may have been the end result of a gradual process. A fire may be made over a body and ritual offerings burnt (such as occurred at Téviec, discussed below); with a shallow grave this could lead to some burning of the body. A number of stages could then be envisioned between this sort of behaviour and full cremation as practised in modern Britain, in which only ashes remain (no prehistoric cremations are this thorough); we would be hard-pressed to say at what stage a 'cremation' was present. Nevertheless, in the time scales that archaeology deals with, certain behaviours *do* appear for the 'first' time, and can be said to be innovative. And in some cases, at least, these innovations may have been sudden and dramatic 'revelations', just as in the modern examples we are more familiar with. Of course, creativity in mortuary behaviour is restricted to neither the place nor the time that I have chosen to focus on here; nevertheless, many of the possibilities for different funerary treatments seem to have been first explored at this time. We must also bear in mind that not all subsequent periods were equally 'creative' in the mortuary sphere.

LIFE IN THE MESOLITHIC

The Mesolithic period of Europe refers to the time between the end of the last glaciation (conventionally taken to be about 10,000 BP) and the adoption of agriculture, or, as Peter Rowley-Conwy (1986) has referred to it, the time 'between cave painters and crop planters'. The end of the period is thus variable across Europe, since agriculture appeared far earlier in the southeast (about 8000 BP or 6800 cal BC) compared to the northwest (about 5200 BP or 4000 cal BC). Hunting of large mammals such as red deer, wild boar, aurochs and roe deer was an important part of the subsistence economy over much of Europe (although to some extent there is a bias in the archaeological record towards the remains of animals, and large animals in particular, which preserve well and are easily identified during excavations). In coastal areas, a variety of fish and marine mammals was exploited. Plants were certainly used, but tend to preserve less well and so are rarely found on archaeological sites of this period. One important exception is hazelnut, ubiquitous on sites in temperate Europe (Zvelebil 1994). Many Mesolithic groups likely exploited their environment in a structured seasonal round, while in some rich areas more permanent occupation may have been possible.

In terms of material culture, a wide variety of stone, bone and antler tools was employed. From waterlogged sites we have the preservation of large dug-out canoes, plant-fibre cordage and nets, and large fish weirs and traps (Andersen 1987). A rather crude form of pottery appears towards the end of the Mesolithic in the circum-Baltic region, possibly in response to the appearance of pottery-using Neolithic cultures to the south. Evidence for artistic expression is for the most part limited to small carvings in bone and antler, and, in the Baltic, amber. Possibly more artistic effort was put into organic materials that have not survived. Over much of Europe, evidence for Mesolithic dwellings has been elusive. This in itself suggests that structures were typically lightly built and ephemeral. Where the remains of dwellings have been found, they tend to be quite small, suggesting that they were occupied by one or two family groups (e.g., Blankholm 1987). The most notable exception to this trend is seen at the site of Lepenski Vir and other sites in the Iron Gates region of the Danube, where a series of structures – possibly ritual in nature rather than domestic – with well-built stone foundations were found (Srejovic 1972).

Mesolithic communities have long been characterised as small, egalitarian and highly mobile, partly based on analogies with highly influential studies undertaken on modern hunter-gatherers such as the !Kung of southern Africa. More recently, this view has been challenged (indeed, much of the impetus for this development came from the discovery of the large and complex cemeteries of southern Scandinavia), and Mesolithic populations are now seen as having encompassed a wider range of social and economic forms. In particular, some coastal and riverine areas appear to have seen more complex organisation, with higher population densities and more permanent occupation of sites.

DEATH IN THE MESOLITHIC

Intentional burial of the dead and accompanying graveside ritual can be traced back to the pre-anatomically modern humans of the Middle Palaeolithic period in Europe and the Near East (Smirnoff 1989; Wymer 1982). The number of burials known in Europe increases dramatically in the Upper Palaeolithic period, and particularly in its later stages. Some graves appear more elaborate relative to others, and it has been suggested that the beginnings of social differentiation can be found at this time (Binford 1968; Harrold 1980; White 1993). One of the most dramatic examples of this is seen at the Aurignacian site of Sungir in Russia, where a complex grave was found with two children lying underneath a headless adult man, which in turn lay underneath the skull of an adult woman placed beside a stone slab in an area stained with red ochre. Thousands of ivory beads and pendants, together with other items, were found with the man and with the children (White 1993). In western Europe, the circulation of deliberately selected parts of the human skeleton seems to be a feature of mortuary behaviour from the Upper Palaeolithic to the Neolithic (Cauwe 1996). The use of ochre and occasional grave offerings suggests increasingly elaborated and articulated concepts of death and the afterworld; more importantly from the perspective of this chapter, the potential of the body and funerary ritual in terms of creating, reinforcing and manipulating social relations appear to be becoming appreciated.

Yet it is not until the Mesolithic that we find evidence for a wide range of mortuary behaviour. Perhaps the most striking change in the Mesolithic is the appearance of comparatively large and well-defined burial grounds (Figure 12.1). It seems that we are fully justified in calling these the first cemeteries (Schulting 1996a), in that they represent conscious use (rather than possibly fortuitous accumulation, such as in the repeatedly visited caves and rockshelters of the Upper Palaeolithic, for example) of a particular place for burial of the dead over a number of generations. Some of the better-known Mesolithic cemeteries include Olenii ostrov in Karelia (Jacobs 1995; O'Shea and Zvelebil 1984), Zvejnieki in Latvia (Zagorska 1993; Zagorska and Larsson 1994; Zagorskis 1973), Skateholm I and II in southern Sweden (Larsson 1984, 1988a,

Figure 12.1 Map showing locations of selected sites dicussed in text

1989; Persson and Persson 1984, 1988), Vedbæk-Bøgebakken and Nederst in Denmark (Albrethsen and Brinch Petersen 1976; Kannegaard and Brinch Petersen 1993), Aveline's Hole in southwest England (Jacobi 1987), Téviec and Hoëdic off the Breton coast of northwest France (Péquart and Péquart 1954; Péquart *et al.* 1937; Schulting 1996b), Lepenski Vir, Vlasac and Padina in the Iron Gates region of the Lower Danube (Prinz 1987; Radovanovic 1996; Srejovic 1972) and the site complexes of the Muge and Sado river valleys in central and southern Portugal (Arnaud 1989; Roche 1972, 1989).

An excellent example of the wide range of contemporaneous Mesolithic mortuary practices is provided by the site of Vlasac in the Iron Gates region of the Lower Danube. Dating to 6850–6260 cal BC (Radovanovic 1996: 367), an early group of burials on the western edge of the site, comprising the remains of some 33 individuals, presented such diverse practices as primary inhumation, cremation, and the secondary burial of crania and longbones. Moreover, these burials were arranged around a structure with stone foundations similar to the habitations of the region, but which is in this case thought to be more directly related to mortuary rituals (Radovanovic 1996: 189–194, figure 4.8). Another important feature of Vlasac, also found at the sites of Lepenski Vir and Hajducka Vodenica, involves the use of some graves for sequential interments, providing some of the earliest European evidence for tombs. The chamber tomb at the rear of a structure at Hajducka Vodenica is the most striking in this regard, containing the remains of a total of 19 individuals, including a series of parallel primary interments as well as secondary burials (Radovanovic 1996: 219–222, figure 4.16).

Some of the most interesting uses of the dead are seen in the Late Mesolithic levels at the site of Lepenski Vir; although there are some problems with the radiocarbon sequence here, as the relevant levels appear to span the range from *c.*7360 to 6800 BP (*c.*6200–5650 cal BC) (Srejovic 1972). Srejovic has suggested the existence of a special relationship between a burial early in the Lepenski Vir sequence and the site's highly distinctive trapezoidal house forms (1972: 117–118). This burial and the habitations, or possibly ritual structures, share a common shape and orientation (Figure 12.2). Srejovic further notes that the position of the skull at the apex of the burial could be paralleled by the placement of an altar or anthropomorphic sculpture in the rear of many of the structures. Indeed, in some cases burials or isolated crania were associated with such sculptures (Srejovic 1972: plate 59). This suggests a powerful metaphor of equivalency between the human body and the structures. There is potentially a further level to the metaphor here, as a striking mountain rising across the river from the site also appears as a trapezoid (Chapman 1993: 101). While this idea is intriguing, it should be approached with some caution; most graves in the Iron Gates Mesolithic are oriented with respect to the river, and seated burials – arguably what is represented by the Lepenski Vir example – are not unknown from other sites (Radovanovic 1996).

Nevertheless, a common underlying theme may be represented by both the human body and the structures in the Iron Gates region; certainly there is an intimate association between the two. A further example of this link can be seen in a number of human mandibles that apparently served as the prototype for a series of triangular stone arrangements built into the paved area around hearths within some of the structures at Lepenski Vir (Figure 12.3; Srejovic 1972). Similar stone triangles have been found at the nearby Mesolithic sites of Padina and Kula (Radovanovic 1996). The interpretation of these elements as symbolic (although possibly in conjunction with a functional purpose) is supported by the treatment given to human and, to a lesser extent, dog mandibles, which are sometimes found as grave inclusions in human burials. Both crania and

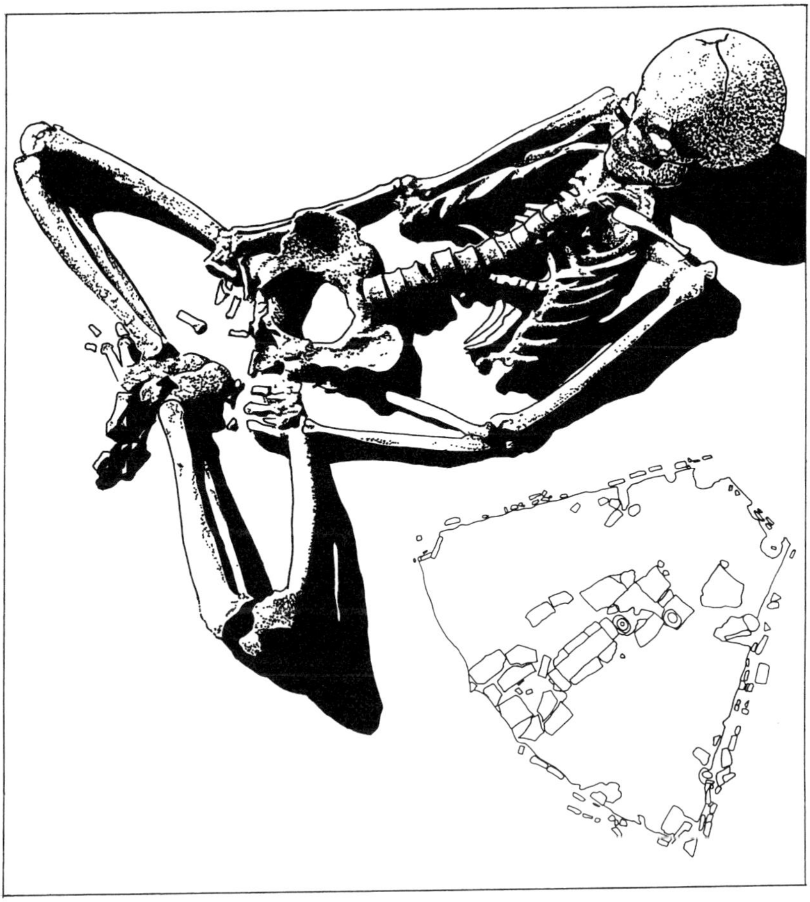

Figure 12.2 Grave 69 from Lepenski Vir, suggesting a possible relationship with the trapezoidal habitation and/or ritual structures at the site (inset)
Source: After Srejovic 1972: plate 56 and fig. 8

mandibles were clearly perceived in the Iron Gates Mesolithic as symbolically powerful elements of the human skeleton. There are numerous examples of secondary burial of these elements, together with some corresponding examples of burials lacking skulls (ibid.). The secondary burial of adult crania – predominantly reserved for older adult males – and mandibles could suggest the use of these elements in ancestor worship (ibid.: 224; Srejovic 1972: 122–124, 1989: 484), similar to what has been widely suggested over much of Europe for the following Neolithic period. The presence of an ancestor within the hearth area, referred to by the crania, mandibles and stone sculptures, may have conferred a degree of ritual authority and seniority within the community.

Figure 12.3 Lepenski Vir I hearth surrounded by triangular stone elements; inset shows human mandible set in paved area around another hearth

Source: After Srejovic 1972: plates 21 and 64

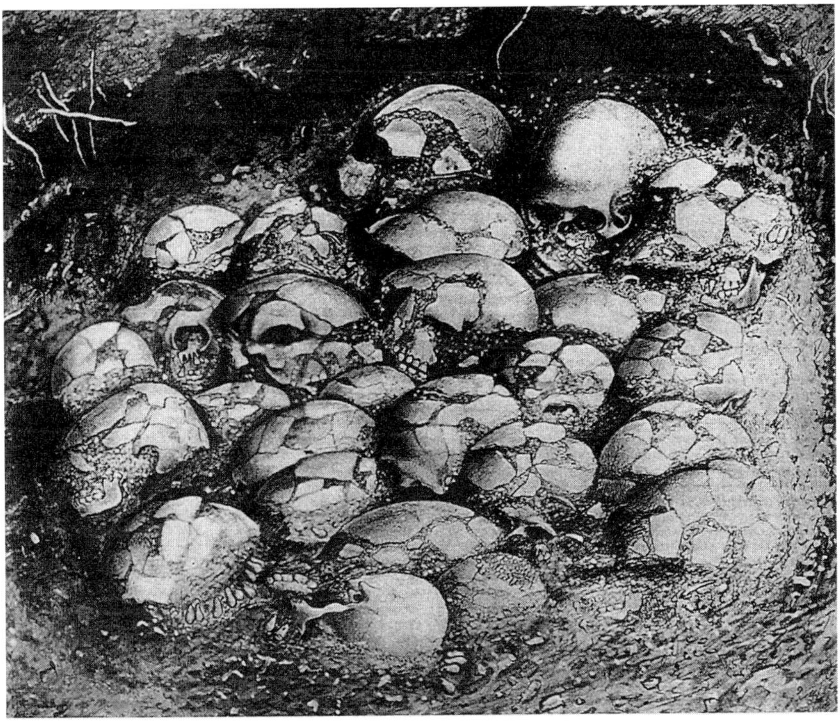

Figure 12.4 Ofnet 'skull nest'; the majority of the 27 skulls are those of women and children
Source: Clark 1967: fig. 128, drawn by Diana Holmes

Another striking burial rite involving the head of the deceased is seen in the 'skull nests' of Germany (Figure 12.4). Ofnet cave produced two of these 'skull nests', one with 27 and the other with six skulls; radiocarbon determinations place the skulls in the range of 7560 to 7360 BP (*c*.6400–6150 cal BC) (Hedges *et al.* 1989). The shallow pits also contained cervical vertebrae and isolated fragments of burned human bone (Newell *et al.* 1979: 155). The presence of cervical vertebrae, together with cut-marks on some of the crania, strongly suggests the decapitation of fleshed bodies. The skulls were arranged in concentric arcs and sprinkled with red ochre. All faced west, towards the cave entrance. Another skull nest is reported from the site of Hohlestein in Württemberg, where three skulls were discovered in a funnel-shaped pit, again coloured with red ochre (Newell *et al.* 1979: 174). All three individuals had been decapitated, as could be seen by damage to cervical vertebrae also found in the pit. Further similar examples could be cited. The use of abundant red ochre and shell ornamentation is a common feature of the skull nests.

The meaning behind these acts is unknown to us, but it clearly involves more than corpse disposal. It seems fairly clear that the heads of many, if not all, of these individuals – the majority of whom were women and children – were intentionally removed while the body was still in a fleshed state. Whether this occurred as a living sacrifice or as treatment given to the recently deceased is impossible to tell, but in either case the importance of the head seems to be implied, perhaps invoking the idea that the head is the receptacle for the spirit or power of the individual.

The cemeteries of southern Scandinavia and Brittany reveal evidence of elaborate mortuary behaviours. The sites of Téviec and Hoëdic in southern Brittany are best known for their evidence of elaborate burial practices, including stone and red deer antler structures, evidence for ceremonial burning and feasting, and abundant and varied gravegoods. Graves at Téviec in particular are associated with stone cists and cairn-like features, often seen as the forerunners of the megalithic tombs that become abundant and elaborate in this area in the Neolithic. There is also possible evidence for feasting associated with the funerary rites, in the form of red deer and wild boar mandibles (presumably symbolic tokens of the feast) placed in light fires made on stone slabs over many of the adult graves at Téviec (Schulting 1996b). Like the Mesolithic sites of the Iron Gates, Téviec and Hoëdic provide some of the first good evidence for the repeated use of constructed sepultures or 'tombs' in Europe; the most recent individuals in a grave were always found in a largely articulated state, while in a number of cases the remains of previous interments had been pushed aside to make room for the later interments (Péquart and Péquart 1954; Péquart *et al.* 1937; Schulting 1996b). It is possible that these represent the tombs of close kin, reflecting the idea that groups together in life should be kept together in death. The single radiocarbon determination from Hoëdic is 6575 BP (*c.*5500–5400 cal BC).

The site complex of Skateholm in coastal southern Sweden (Scania) has provided a large series of burials, both adjacent to and partly overlying occupation layers (Larsson 1988a, 1989). A series of seven radiocarbon estimates places them in the range 6300 to 5900 BP (*c.*5250–4800 cal BC). The burials are highly variable in terms of body position and orientation, and both inhumations and cremations are present. Here we find early evidence for the use of above-ground structures associated both directly and indirectly with mortuary rites. The placement of postholes around Grave 11 at Skateholm I, containing the widely scattered and burnt bones of an adult male together with diverse animal remains, suggests the presence of a wooden structure presumably associated with the funerary ritual (Larsson 1989: 372). Another structure apparently associated with mortuary-related activities at the site was found near the edge of the cemetery. A centrally placed structure at Skateholm II has been interpreted as a ceremonial building involved in mortuary activities at the site (Larsson 1988b). The feature, measuring some 4 by 4 m, was defined by a border consisting of an intense layer of red ochre, within which were found distinct concentrations of flint and animal bone, including species that were rare or absent in the occupation midden. Several postholes were found within the confines of the feature. The placement of the feature is also of interest, as it was situated on the highest point of ground and marked an interruption in the line of graves along a ridge.

One of the most fascinating aspects of Skateholm concerns the burial not of people, but of dogs. A total of some ten clearly intentional dog burials were found at Skateholm I and Skateholm II, while the remains of a further seven dogs were found associated with human burials (Larsson 1990). Dog burials have also been found in Denmark (Brinch Petersen *et al.* 1993; Kannegaard Nielsen and Brinch Petersen 1993). The dog of Grave XXI at Skateholm II was found lying in a grave pit together with goods more commonly associated with adult human burials – flint blades, red deer antler and a decorated antler hammer (Figure 12.5). The grave was also surrounded with red ochre. As Lars Larsson (1989, 1990, 1993) has suggested, it seems clear that this extends beyond the treatment given

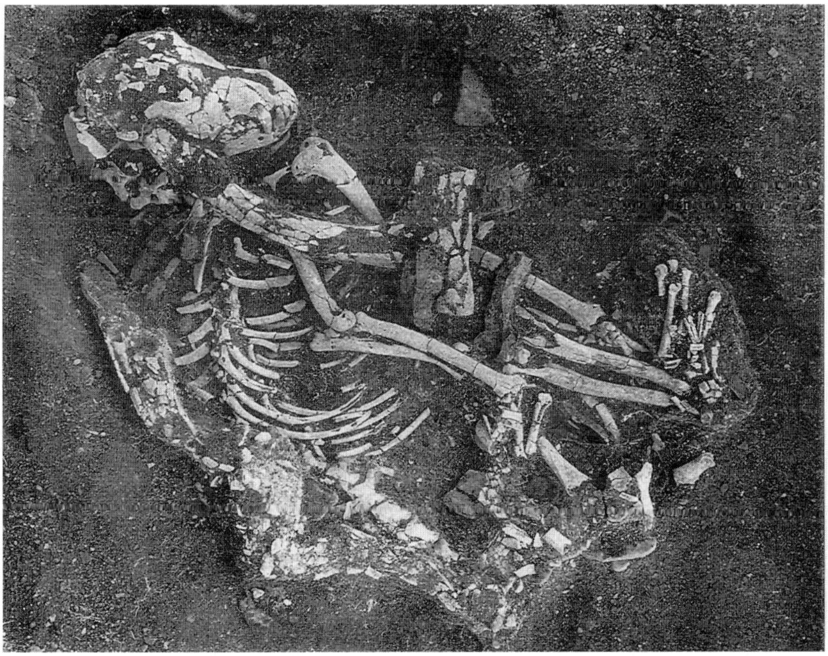

Figure 12.5 Grave XXI at Skateholm II, showing dog with ornamented antler axe and flint blades
Source: Larsson 1995: fig. 7

to a beloved pet or favoured hunting dog; there appears to be the suggestion of an equivalency between animal and human. Pamela Amoss (1984) has commented on the ambiguous position of the dog among the Native peoples of the Northwest Coast of North America, suggesting that dogs served as symbols of mediation between the domains of man and beast. This ability to recognise – or rather create – analogies of this nature is intimately linked to creativity (Boden 1990). I suggest that the dogs given such treatment in the Mesolithic of southern Scandinavia may have become involved in social relationships between people. Perhaps particularly valued dogs were exchanged between communities to cement alliances. Among the Bella Coola of the Northwest Coast, for example, there is a case in which a hereditary title was passed to a dog (Figure 12.6). The dog in fact became a 'chief' and was entitled to be addressed as one; a potlatch was held to validate the dog's status, just as would be done for a human claimant.

> In November, 1923, the writer, while walking through the Bella Coola village, was bitten by a dog. He was told that the animal belonged to a certain old woman. When next he saw the owner he asked her whether she knew that her dog attacked passers-by. 'Yes', she answered in a matter of fact tone, 'He is a chief and can do what he likes.' Further questioning elicited the information that the old woman and her husband, the childless last survivors of an ancestral family, had transmitted some of their ancestral names to their dog, and had distributed presents to validate the bestowal. Consequently, the animal is a chief and can do as he likes. Moreover, a

Figure 12.6 A Bella Coola couple with their dogs; the one on the left is a chief
Source: University of Toronto Archives, Thomas F. McIlwraith papers, acc.#B79–0054, Box 3

> number of people have returned the presents which they received at the time the dog was made a chief, and this money is kept by the woman and her husband, to be used when they perform the rite of mourning for the dog at death.
>
> (McIlwraith 1992: 174–175)

While it is true that this occurred in a context of massive depopulation caused by introduced diseases, so that no suitable human heir could be found, it does serve to make the point that dogs have been used in complex social roles. Were dogs being used at Skateholm in order to manipulate burial ritual and power relations? Perhaps Larsson's (1989) characterisation of the occupant of Grave XXI as a 'Big Dog' is not so wide of the mark.

Cremations are rare but widespread in the Mesolithic, with examples known from southern Sweden, Denmark, Holland, France, Poland and the Iron Gates (Kannegaard Nielsen and Brinch Petersen 1993; Radovanovic 1996). An example is provided by the site of Gøngehusvej 7 near Vedbæk. One of the two cremations found here contained the remains of five individuals, representing ages ranging from adult to newborn. The bodies had apparently been dismembered and defleshed prior to cremation (Brinch Petersen *et al.* 1993). The meaning behind cremation and the context in which it was chosen from among the available burial options (both of which would certainly have varied through time and space) are not clear, but the display qualities of a large fire and the far more rapid transformation of the deceased certainly offer possibilities for the creative use of symbolism and theatrics.

The submerged site of Møllegabet II off the Danish coast reveals yet another kind of Mesolithic burial practice. Here the 2.5-m-long stern section of a dug-out canoe was found held down in position with pointed stakes at either end. The remains of a young adult man were found scattered around and inside the boat, together with various flint, bone, antler and wooden tools (Grøn and Skaarup 1991). It appears that the body had been placed in the canoe, which was then intentionally sunk and staked down in shallow water. This took place around 5910 BP (*c.*4800 cal BC). Perhaps this is one of the first times that a boat burial was used as a metaphor for a journey that the dead must take. Over 1,000 years later, a very similar rite was enacted in the Neolithic boat burial at Øgårde (Christensen 1990). Was the idea 'rediscovered' at this time? Or is there a continuity in the practice that we cannot see simply because of the chance nature of archaeological preservation and recovery?

The position of Romeiras, with its high concentration of burials, is unique in the Sado River valley of south-central Portugal. It is located across a steep ravine from the large midden of Cabeço do Pez, and it is suggested that its inhabitants were using Romeiras for burial (Arnaud 1989). No radiocarbon determinations are available from Romeiras itself, but those from Cabeço do Pez range from 6700 to 5500 BP (*c.*5550–4350 cal BC).

The positioning of the bodies at Romeiras is unusual, with the individuals placed radially in a semi-circle with the opening facing the river (Arnaud 1989: 621). When graves make reference to one another in this fashion (as opposed to the more common cardinal orientations seen at many other times and in many places), it appears that the meaning of the arrangement is carried beyond each individual burial to the whole; that is, there is an overall plan, the implementation of which necessitates passing over other burial options. In such a situation there appears to be little room for innovation involving individual burials. But Romeiras as a whole does seem to suggest that some group or individual within the community acted at the very least as a coordinator for the disposal aspect of the funerary rites for a number of community members. This in itself could be seen as a creative act involving control and manipulation of ritual practices associated with the dead.

As well as the more unusual and unique examples I have focused on thus far, there is of course also a large number of burials that share many traits. The following example is particularly striking. The most recent discovery from Gøngehusvej 7 on Zealand is an inhumation containing two individuals, an adult woman together with a 3-year-old child. Ochre covered the child completely, but was more patterned on the adult, suggesting placement of ochre around a skin garment (Brinch Petersen *et al.* 1993: figs 3 and 4); this is supported by the presence of roe deer phalanges on the woman's torso. Both individuals wore elaborate ornaments consisting of pendants made of the teeth of red and roe deer, boar, elk, aurochs and bear, together with bird beaks, roe deer hooves, and bone and stone pendants. Other gravegoods included flint knives, a bone point and two bone skinning knives. The most interesting feature of this burial consists in its remarkable similarity to Grave 8 at Vedbæk-Bøgebakken (compare Brinch Petersen *et al.* 1993: fig. 4 to Albrethsen and Brinch Petersen 1976: figs 9 and 10). Both are graves of adult women with a child or newborn on their right hip; in both cases the child has a flint blade over its lower abdomen (perhaps held in a pouch there?); and in both cases there is elaborate tooth ornamentation and abundant red ochre. The two sites are not far from one another, and are roughly contemporary. The near-identical treatment of woman and child in both cases speaks of very formal funerary behaviour and shared symbolism. Thus there was always the option of following a more normative mortuary behaviour, or innovating in a subtle fashion, such as the positioning of the hands for example.

The creative use of the symbolic potency of the corpse is not limited to behaviours surrounding the different forms of its disposal; there is also evidence for the circulation of parts of the skeleton amongst the living. The case of the use of mandibles in stone settings at Lepenski Vir, Padina and Kula has already been mentioned. Drilled pendants of human incisors have been found with some Danish Mesolithic burials (Albrethsen and

Brinch Petersen 1976). A child's rib in a grave at Téviec in Brittany was found to be incised with series of parallel lines on three surfaces (Marshack 1972; Schulting 1996b). This could not have been done with the body in a fleshed state. This then implies the removal of the rib after decomposition of the corpse, the making of the incisions, and the eventual reburial of the element. Cauwe (1996) documents a number of additional examples in France, one of the more striking of which involves deer vertebrae set into the orbits of a human cranium at Mas d'Azil. At both of the Early Mesolithic sites of Grotte Margaux and Abri des Autours in Belgium, a cranium was intentionally smashed and placed beside or over a collective burial (Cauwe 1996). Rozoy (1978: 1167) has noted that isolated human remains in the French Mesolithic involve, in all but one case, the skull or cranial fragments. A similar bias towards skulls, although less extreme, is also seen in the preceding Palaeolithic period (Wymer 1982: 169–171), and continues with the mortuary monuments of the Neolithic (Cauwe 1996). Rozoy sees this as a deliberate choice involving ritual and identification of the head as the seat of personhood, recalling the interpretation offered above for Ofnet. The special treatment accorded to the skull (both cranium and mandible) is especially conspicuous in the Iron Gates Mesolithic sites. There are further examples of skeletons, otherwise well preserved, missing certain elements (Larsson 1993). It is unlikely that post-depositional disturbance could account for all of these cases. It seems logical to assume that, as with many societies in the ethnographic literature from around the world, bones were selectively removed and put to various uses. The dead were buried but not forgotten, and could still be enlisted to aid the living.

DISCUSSION

Clearly, much of what went on during funerary rites in the European Mesolithic is inaccessible, but the glimpses that do remain suffice to show that many novel approaches to the treatment of the dead were explored. The first sepultures appeared, and possibly also the first cremations and boat burials. Specialised structures were erected within cemeteries, such as at Vlasac in the Iron Gates and Skateholm II in southern Sweden, demonstrating the expanding sphere of activities surrounding death and burial. Selected skeletal elements were in some cases circulated and modified. Both the unusual position of the mandibles around hearths at Lepenski Vir and the dog burials at Skateholm demonstrate, I believe, the creative use of analogy. We can also see how these developments were embedded within particular social circumstances. Permanent or semi-permanent occupation has been proposed for many of the settlements associated with cemeteries (Larsson 1984; Lentacker 1986; Radovanovic 1996; Rowley-Conwy 1981, 1983; Schulting 1996b; Srejovic 1972). Increasing population and social inequality at this time may have provided an arena in which establishing and maintaining rights to land and resources became important, and use of the ancestral dead was one means of achieving this end (cf. Larsson 1984, 1995).

Julian Thomas (1996) has recently re-emphasised the potential cultural dynamism and hybridity occurring along the periphery of the Early Neolithic Linearbankeramik (LBK) settlement in northwest Europe. He points to the selective modification and incorporation of certain aspects of novel technology by indigenous Mesolithic populations. One end result of this process is seen in the widespread adoption of certain forms of mortuary monuments in the Neolithic, particularly earthen long barrows and megalithic chambered tombs. The distribution of these along the Atlantic façade that saw some of the most intense Mesolithic developments has not gone unremarked (Renfrew 1976; Sherratt 1990, 1995). Obviously the idea of the long barrow and chambered tomb originated somewhere, and from thence became widespread, but, as Thomas has argued, their meaning need not have remained the same over the entire area. While the form is largely provided by tradition, the way in which it is interpreted in a particular instance can involve a creative process (this is not to say that it always does, or that we will be able to tell when a creative use is being made of a traditional form).

What seems at first a distant analogy can be made with the adoption of Christianity by Native peoples across western North America in the early nineteenth century. Christianity and its trappings, which included an entirely different burial position and orientation from that practised in pre-contact times (Sprague 1967), was embroiled in social and political contexts. Some high-ranking individuals were among the first to 'convert' (though it is questionable how deep this conversion went at first) in order to gain an advantage in the highly profitable trade relations with Euro-americans. But there were also more traditional elements in society that resisted conversion, creating factional tensions. And of course hybrid religions also appeared, such as among the followers of the Ghost Dance. Most importantly, some of these developments occurred in contexts in which the missionaries themselves had not yet made an appearance. This more than anything else demonstrates the active and complex nature of processes of cultural interaction along 'frontiers'. While not wishing to labour the point, and acknowledging its limitations, I suggest that there may be conceptual parallels between the adoption of Christianity by Native Americans and the widespread adoption of long barrows and chambered tombs in the Early Neolithic of western Europe. In both cases, the new behaviours can be viewed as creative responses to changing conditions, and, more specifically, new opportunities that individuals and groups could manipulate to their advantage.

The focus on the specific sites and areas we have been discussing – particularly southern Scandinavia and the Lower Danube – seems to carry with it the implication that these areas harboured more 'creativity' than other areas at this time. Because we are dealing with a period removed from us by some 6,000 years or more, we can attribute much of our focus to accidents of preservation and recovery. Britain, for example, does not enter into the discussion for the simple reason that there is very little bone preservation from the Mesolithic. But even if we had evidence of mortuary practices from these 'blank' areas, there is no guarantee that they would add much of interest to the present discussion. I attribute this not to a lack of creativity on the part of the people in these areas, who indeed may have showed creativity in other ways, but to differing social contexts. Areas that see the innovative burial practices I have mentioned are often pointed to as areas evidencing increasing social complexity (on the basis of criteria mentioned above and below). It is this social context that provides the incentive for innovation in the sphere of mortuary behaviour.

Burials among 'simple' hunter-gatherer societies known ethnographically for the most part *are* simple and, more to the point here, homogeneous affairs (Woodburn 1982; see also references in Hofman 1986). Cemeteries are practically unknown; the body may be buried in a shallow pit, or simply covered with brush and abandoned. No doubt these practices characterise much of early prehistory around the world. But among what have become known as complex hunter-gatherers (Price and Brown 1985), best typified by the Native peoples of the Northwest Coast of North America, mortuary behaviour is often complex and diverse. Once the more egalitarian principles typical of many hunter-gatherer societies are eroded, there comes into existence an incentive to enlist funerary rites for display and competitive behaviour. Material culture and behaviour associated with mortuary rites merge to help create as well as reflect (whether 'truthfully' or not is beside the point in this context) new, more differentiated social roles. The visible presence of the ancestors in the form of marked graves and formal cemeteries can serve as markers of rights of access to important natural or social resources. Ritual power can accumulate in the hands of those who specialise in the appeasing of the dead, or in facilitating communication between the dead and the living. Funerals, and especially the feasts often associated with them, serve as occasions of conspicuous display where efforts can be made to attract followers and create new alliances by advertising the success of the deceased's kin group. Thus it is the ability to make use of innovation that calls forth innovation – to paraphrase, 'Opportunity is the mother of invention.'

This leads us then to a consideration of what group within these societies would be the most likely innovators. It has frequently been suggested that innovations are often first accepted by the socio-economic elite as a means of distinguishing themselves and gaining advantage over their peers (Cannon 1989; Fallers 1973; Randsborg 1982; Shennan 1982). I should add that by the term 'elite' here I refer to nothing more than that subgroup that holds most of the power within a given society (see Blau 1977). By contrast with innovation, the process of emulation refers to the copying of elite behaviours by lower social subgroups in an effort to acquire the prestige associated with elite behaviour (Miller 1982). Thus we can propose that innovation in mortuary behaviour generally took place first among those social groups already invested with, or vying for, power, as they stood the most to gain. But the social system was not yet a static one with clearly established hierarchies – in such a case the elite may become the more conservative group, as they stand to lose the most. And even in a situation of competing groups, it is often the particular circumstances at the time that will determine the extent to which a particularly elaborate or innovative mortuary display would be perceived by those hosting the event as advantageous (e.g. Metcalf 1981).

What was it that stood to be gained? We know from the ethnographic literature that there are many reasons for political manoeuvring even in small-scale societies. Competition can occur both within and between communities for access to prime resources, marriage partners, trading partners and alliances. Mortuary behaviour then becomes one aspect of this process. From the European Mesolithic we have evidence for long-distance trade in important raw materials, such as particular types of stone for the manufacture of tools (Gendel 1982; Palmer 1970; Schild 1976; Wickham-Jones 1990). There are hints of investment in stationary facilities such as large fish weirs and nets that could have required the coordinated efforts of a labour pool extending beyond the local kin group (Enghoff 1995; Pedersen 1995). And there is evidence of considerable inter-personal violence (Balakin and Nuzhnyi 1995; Bennike 1985; Grünberg 1996; Newell *et al.* 1979; Péquart *et al.* 1937; Persson and Persson 1984, 1988; Radovanovic 1996), possibly suggesting the need for alliances.

It is in this context that much innovative mortuary behaviour appears to have occurred. The rites surrounding funerals presented opportunities to create new or emphasise existing social roles, to create and pay debts, and to attract and bind followers. Indeed, one of the more 'creative' acts surrounding the funeral that would not be directly visible might consist of efforts made by the surviving kin to secure the participation of a wide segment of the community. In small-scale societies, rituals such as those surrounding death are often employed as a kind of legal contract (e.g., the Northwest Coast potlatch). The more memorable the event, the more binding the contract, and the more successful it would be judged. An excellent ethnographic example of this process is presented by the *-murina* cycle of mortuary feasts held by the East Solomon Islanders, where innovation in reburial rites is strongly associated with status and conducted with the express goal of having the event discussed and remembered (Davenport 1986). This is not to say that the organisers of each funeral strove to orchestrate a unique and unprecedented event. There were always options – to follow what had previously been successful, or to innovate in some small way that would not be visible in the archaeological record. And the mortuary rite must still meet cultural notions of propriety; innovation must occur within the bounds of, or at least not too far outside of, an accepted grammar. There were risks as well as benefits to pushing the limits of this grammar. Thus attempts by lower-ranking lineages among the Baruya of Papua New Guinea to renegotiate vested ritual authority by creating their own ritually powerful objects were met with scorn and ridicule by the rest of the community – the attempts failed dismally (Godelier 1986). As Chris Tilley (1994:30) has recently commented, only the high-ranking or wealthy person is likely to be daring enough to invent a path or plant a relationship not established before.

The aspect of sociocultural complexity that I have stressed in this chapter is inequality. The other facet of complexity is heterogeneity, which may be thought of as the number of non-hierarchically ordered social roles (Blau 1977). It is not my intention to downplay this aspect; certainly the increasing (in general) number of possible social roles in prehistory can be expected to be reflected in mortuary treatment (as can the circumstances surrounding death) (Binford 1971; Kingsley 1985). But heterogeneity by definition is not associated with inequality (although in practice the two may be inseparably linked, as with ethnicity or gender for example), and in this sense does not provide the same incentive for innovation in the use of material culture and behaviour that I see as offering the best interpretative framework for exploring the concept of 'creativity' in the mortuary sphere.

ACKNOWLEDGEMENTS

I would like to thank Steve Mithen for the opportunity to participate in this volume, and for his comments on a draft manuscript. Thanks also to Richard Bradley, Nyree Finlay, Brian Hayden and Joanna Ostapkowicz for their comments and criticisms on an earlier draft of this chapter – they are in no way to blame for any shortcomings in the final product. Figures 12.2 and 12.3 were drawn by Joanna Ostapkowicz. I am grateful to Thames and Hudson, Lars Larsson and Tom McIlwraith for permission to reproduce Figures 12.4, 12.5 and 12.6, respectively. Finally, I would like to acknowledge the support of the British Office of the Commonwealth Association and of the Social Sciences and Humanities Research Council of Canada.

REFERENCES

Albrethsen, S.E. and Brinch Petersen, E. (1976) 'Excavation of a Mesolithic cemetery at Vedbæk, Denmark.' *Acta Archaeologica* 47: 1–28.

Amoss, P. (1984) 'A little more than kin, and less than kind: the ambiguous

Northwest Coast dog.' In J. Miller and C.M. Eastman (eds) *The Tsimshian and Their Neighbours of the North Pacific Coast*, pp. 292–305. Seattle: University of Washington.

Andersen, S.H. (1987) 'Mesolithic dug-outs and paddles from Tybrind Vig, Denmark.' *Acta Archaeologica* 57: 87–106.

Arnaud, J.E.M. (1989) 'The Mesolithic communities of the Sado Valley, Portugal, in their ecological setting.' In C. Bonsall (ed.) *The Mesolithic in Europe*, pp. 614–631. Edinburgh: John Donald.

Balakin, S. and Nuzhnyi, D. (1995) 'The origin of graveyards: the influence of landscape elements on social and ideological changes in prehistoric communities.' *European Prehistory* 7: 191–202.

Bennike, P. (1985) *Palaeopathology of Danish Skeletons*. Copenhagen: Akademisk Forlag.

Binford, L.R. (1971) 'Mortuary practices: their study and their potential.' In J.A. Brown (ed.) *Approaches to the Social Dimensions of Mortuary Practices*, pp. 6–29. New York: Memoirs of the Society for American Archaeology.

Binford, S.R. (1968) 'A structural comparison of disposal of the dead in the Mousterian and the Upper Palaeolithic.' *Southwestern Journal of Anthropology* 24: 139–154.

Blankholm, H.P. (1987) 'Maglemosian hutfloors: an analysis of the dwelling unit, social unit and intra-site behavioural patterns in Early Mesolithic southern Scandinavia.' In P. Rowley-Conwy, M. Zvelebil and H.P. Blankholm (eds) *Mesolithic Northwest Europe: Recent Trends*, pp. 109–120. Sheffield: University of Sheffield Press.

Blau, P.M. (1977) *Inequality and Heterogeneity: A Primitive Theory of Social Structure*. New York: The Free Press.

Boden, M.A. (1990) *The Creative Mind: Myths and Mechanisms*. London: Cardinal.

Brinch Petersen, E., Alexandersen, V. and Meiklejohn, C. (1993) 'Vedbæk, graven midt i byen.' *Nationalmuseets Arbejdsmark*: 61–69.

Cannon, A. (1989) 'The historical dimension in mortuary expressions of status and sentiment.' *Current Anthropology* 30(4): 437–458.

Cauwe, N. (1996) 'Les sépultures collectives dans l'espace et le temps.' *Bulletin de la Société Préhistorique Française* 93: 342–352.

Chapman, J. (1993) 'Social power in the Iron Gates Mesolithic'. In J. Chapman and P. Dolukhanov (eds) *Cultural Transformations and Interactions in Eastern Europe*, pp. 71–121. Aldershot: Avebury.

Christensen, C. (1990) 'Stone age dug-out boats in Denmark: occurrence, age, form and reconstruction.' In D. Robinson (ed.) *Experimentation and Reconstruction in Environmental Archaeology*, pp. 119–141. Oxford: Oxbow Books.

Clark, J.D.G. (1967) *The Stone Age Hunters*. London: Thames & Hudson.

Davenport, W.H. (1986) 'Two kinds of value in the Eastern Solomon Islands.' In A. Appadurai (ed.) *The Social Life of Things: Commodities in Cultural Perspective*, pp. 95–109. Cambridge: Cambridge University Press.

Douglas, M. (1996) *Natural Symbols*. London: Routledge.

Enghoff, I.B. (1995) 'Fishing in Denmark in the Mesolithic period.' In A. Fischer (ed.) *Man and Sea in the Mesolithic*, pp. 67–74. Oxford: Oxbow Books.

Fallers, L.A. (1973) *Inequality: Social Stratification Reconsidered*. Chicago: University of Chicago Press.

Geary, P. (1986) 'Sacred commodities: the circulation of medieval relics.' In

A. Appadurai (ed.) *The Social Life of Things: Commodities in Cultural Perspective*, pp. 169–191. Cambridge: Cambridge University Press.

Gendel, P.A. (1982) 'The distribution and utilization of Wommersom quartzite during the Mesolithic.' In A. Gob and F. Spier (eds) *Le Mésolithique entre Rhin et Meuse*, pp. 21–50. Luxembourg: Société Préhistorique Luxembourg.

Godelier, M. (1986) *The Making of Great Men: Male Domination and Power Among the New Guinea Baruya*. Cambridge: Cambridge University Press.

Grøn, O. and Skaarup, J. (1991) 'Møllegabet II – a submerged Mesolithic site and a "boat burial" from Æro.' *Journal of Danish Archaeology* 10: 38–50.

Grünberg, J.M. (1996) 'Mesolithic burial practices in Europe: selection and inequality in the treatment of the dead.' Paper presented at the 13th International Congress of Prehistoric and Protohistoric Sciences, Forlì, Italy.

Harrold, F. (1980) 'A comparative analysis of Eurasian Paleolithic burials.' *World Archaeology* 12: 195–211.

Hedges, R.E.M., Housley, R.A., Law, I.A. and Bronk Ramsey, C. (1989) 'Radiocarbon dates from the Oxford AMS system: *Archaeometry* Datelist 9.' *Archaeometry* 31(2): 207–234.

Hertz, R. (1907) 'Contribution à une étude sur la représentation de la mort.' *Anneé Sociologique* 10: 48–137.

Hofman, J.L. (1986) *Hunter-Gatherer Mortuary Variability: Toward an Explanatory Model*. Ann Arbor: University Microfilms.

Jacobi, R.M. (1987) 'Misanthropic miscellany: musings on British early Flandrian archaeology and other flights of fancy.' In P. Rowley-Conwy, M. Zvelebil and H.P. Blankholm (eds) *Mesolithic Northwest Europe: Recent Trends*, pp. 163–168. Sheffield: University of Sheffield Press.

Jacobs, K. (1995) 'Returning to Oleni' Ostrov: social, economic, and skeletal dimensions of a boreal forest mesolithic cemetery.' *Journal of Anthropological Archaeology* 14(4): 359–403.

Kannegaard Nielsen, E. and Brinch Petersen, E. (1993) 'Burials, people and dogs.' In S. Hvass and B. Storgaard (eds) *Digging into the Past. 25 Years of Archaeology in Denmark*, pp. 76–81. Århus: Århus University Press.

Kingsley, R.G. (1985) *Kin Groups and Mortuary Practices: Ethnographic Implications for Archaeology*. Ann Arbor: University Microfilms.

Larsson, L. (1984) 'The Skateholm Project: a late Mesolithic settlement and cemetery complex at a southern Sweden bay.' *Meddelanden från Lunds Universitets Historiska Museum 1983–1984* 5: 5–38.

Larsson, L. (1988a) *The Skateholm Project. I. Man and Environment*. Acta Regiae Societatis Humaniorum Litterarum Ludensis LXXIX, Stockholm: Almqvist and Wiksell International.

Larsson, L. (1988b) 'A construction for ceremonial activities from the Late Mesolithic.' *Meddelanden från Lunds Universitets Historika Museum 1987–1988* 7: 5–18.

Larsson, L. (1989) 'Big dog and poor man. Mortuary practices in Mesolithic societies of southern Sweden.' In T.B. Larsson and H. Lundmark (eds) *Approaches to Swedish Prehistory. A Spectrum of Problems and Perspectives in Contemporary Research*, pp. 211–223. Oxford: BAR International Series 500.

Larsson, L. (1990) 'Dogs in fraction – symbols in action.' In P.M. Vermeersch and P. van Peer (eds) *Contributions to the Mesolithic in Europe*, pp. 153–160. Leuven: Leuven University Press.

Larsson, L. (1993) 'The Skateholm Project: Late Mesolithic coastal settlement in southern Sweden.' In P. Bogucki (ed.) *Case Studies in European Prehistory*, pp. 31–62. Boca Raton, Florida: CRC Press.

Larsson, L. (1995) 'Man and sea in southern Scandinavia during the Late Mesolithic: the role of cemeteries in the view of society.' In A. Fischer (ed.) *Man and Sea in the Mesolithic*, pp. 95–104. Oxford: Oxbow Books.

Lentacker, A. (1986) 'Preliminary results of the fauna of Cabeço da Amoreira and Cabeço da Arruda.' *Trabalhos de Antropologia e Etnologia* 26: 9–26.

McIlwraith, T.F. (1992) *The Bella Coola Indians*. 2 volumes (original 1942). Toronto: University of Toronto Press.

Marshack, A. (1972) *The Roots of Civilization*. New York: McGraw-Hill.

Metcalf, P. (1981) 'Meaning and materialism: the ritual economy of death.' *Man* (n.s.) 16: 563–578.

Metcalf, P. and Huntington, R. (1991) *Celebrations of Death: The Anthropology of Mortuary Ritual*. Cambridge: Cambridge University Press.

Miller, D. (1982) 'Structures and strategies: an aspect of the relationship between social hierarchy and cultural change.' In I. Hodder (ed.) *Symbolic and Structural Archaeology*, pp. 89–98. Cambridge: Cambridge University Press.

Newell, R.R., Constandse-Westermann, T.S. and Meiklejohn, C. (1979) 'The skeletal remains of Mesolithic man in western Europe: an evaluative catalogue.' *Journal of Human Evolution* 8(1): 1–228.

O'Shea, J. and Zvelebil, M. (1984) 'Oleneostrovski Mogilnik: reconstructing the social and economic organisation of prehistoric foragers in northern Russia.' *Journal of Anthropological Archaeology* 3(1): 1–40.

Palmer, S. (1970) 'The Stone Age industries of the Isle of Portland, Dorset, and the utilisation of Portland chert as artifact material in southern England.' *Proceedings of the Prehistoric Society* 36: 82–115.

Pedersen, L. (1995) '7000 years of fishing: stationary fishing structures in the Mesolithic and afterwards.' In A. Fischer (ed.) *Man and Sea in the Mesolithic*, pp. 75–86. Oxford: Oxbow Books.

Péquart, M. and Péquart, S.-J. (1954) *Hoëdic, Deuxième Station-Nécropole du Mésolithique Côtier Armoricain*. Anvers: De Sikkel.

Péquart, M., Péquart, S.-J., Boule, M. and Vallois, H. (1937) *Téviec, Station-Nécropole du Mésolithique du Morbihan*. Paris: Archives de L'Institut de Paléontologie Humaine XVIII.

Persson, O. and Persson, E. (1984) *Anthropological Report on the Mesolithic Graves from Skateholm, Southern Sweden. I. Excavation Seasons 1980–1982*. Institute of Archaeology, Report Series no. 21, Lund.

Persson, O. and Persson, E. (1988) 'Anthropological report concerning the interred Mesolithic populations from Skateholm, southern Sweden. Excavation seasons 1983–1984.' In L. Larsson (ed.) *The Skateholm Project. I. Man and Environment*, pp. 89–105. Acta Regiae Societatis Humaniorum Litterarum Ludensis LXXIX, Stockholm: Almqvist & Wiksell International.

Price, T.D. and Brown, J.A. (eds) (1985) *Prehistoric Hunter-Gatherers: The Emergence of Cultural Complexity*. New York: Academic Press.

Prinz, B. (1987) *Mesolithic Adaptions on the Lower Danube: Vlasac and the Iron Gates Gorge*. Oxford: BAR International Series 330.

Radovanovic, I. (1996) *The Iron Gates Mesolithic*. Ann Arbor: International Monographs in Prehistory.

Randsborg, K. (1982) 'Rank, rights and resources – an archaeological perspective from Denmark.' In C. Renfrew and S. Shennan (eds) *Ranking, Resource and Exchange: Aspects of the Archaeology of Early European Society*, pp. 132–139. Cambridge: Cambridge University Press.

Renfrew, C. (1976) 'Megaliths, territories and populations.' In S.J. De Laet (ed.) *Acculturation and Continuity in Atlantic Europe*, pp. 198–220. Bruges: De Tempel.

Renfrew, C. (1978) 'The anatomy of innovation.' In D. Green, C. Haselgrove and M. Spriggs (eds) *Social Organisation and Settlement*, pp. 89–117. Oxford: BAR International Series 47 (i).

Roche, J. (1972) *Le Gisement Mésolithique de Moita do Sebastião, Muge, Portugal. I: Archéologie*. Lisbon: Instituto de Alta Cultura.

Roche, J. (1989) 'Spatial organization in the Mesolithic sites of Muge, Portugal.' In C. Bonsall (ed.) *The Mesolithic in Europe*, pp. 607–613. Edinburgh: John Donald.

Rowley-Conwy, P. (1981) 'Mesolithic Danish bacon: permanent and temporary sites in the Danish Mesolithic.' In A. Sheridan and G. Bailey (eds) *Economic Archaeology*, pp. 51–55. Oxford: BAR International Series S96.

Rowley-Conwy, P. (1983) 'Sedentary hunters: the Ertebølle example.' In G. Bailey (ed.) *Hunter Gatherer Economy in Prehistory*, pp. 111–126. Cambridge: Cambridge University Press.

Rowley-Conwy, P. (1986) 'Between cave painters and crop planters: aspects of the temperate European Mesolithic.' In M. Zvelebil (ed.) *Hunters in Transition*, pp. 17–32. Cambridge: Cambridge University Press.

Rozoy, J.-G. (1978) *Les Derniers Chasseurs. L'Épipaléolithique en France et en Belgique. Essai de Synthèse*. Charleville: Bulletin de la Société Archéologique Champenoise, J.-G. Rozoy.

Schild, R. (1976) 'Flint mining and trade in Polish prehistory as seen from the perspective of the chocolate flint of central Poland. A second approach.' *Acta Archaeologica Carpatica* 16: 147–170.

Schulting, R.J. (1996a) '"A Place Set Apart . . .": The question of Mesolithic cemeteries.' Paper presented at the 61st annual meeting of the Society for American Archaeology, New Orleans.

Schulting, R.J. (1996b) 'Antlers, bone pins and flint blades: the Mesolithic cemeteries of Téviec and Hoëdic, Brittany.' *Antiquity* 70: 335–350.

Shennan, S.J. (1982) 'Ideology, change and the European Early Bronze Age.' In I. Hodder (ed.) *Symbolic and Structural Archaeology*, pp. 155–161. Cambridge: Cambridge University Press.

Sherratt, A. (1990) 'The genesis of megaliths: monumentality, ethnicity and social complexity in Neolithic north-west Europe.' *World Archaeology* 22(2): 147–167.

Sherratt, A. (1995) 'Instruments of conversion? The role of megaliths in the Mesolithic/Neolithic transition in north-west Europe.' *Oxford Journal of Archaeology* 14(6): 245–260.

Simonton, D.K. (1994) 'Individual differences, developmental changes, and social context.' *Behavioural and Brain Sciences* 17: 552–553.

Smirnoff, Y. (1989) 'Intentional human burial: Middle Palaeolithic late glaciation beginnings.' *Journal of World Prehistory* 3: 199–233.

Sprague, R. (1967) *Aboriginal Burial Practices in the Plateau Region of North America.*

Ann Arbor: University Microfilms.

Srejovic, D. (1972) *Europe's First Monumental Sculpture: New Discoveries at Lepenski Vir*. London: Thames & Hudson.

Srejovic, D. (1989) 'The Mesolithic of Serbia and Montenegro.' In C. Bonsall (ed.) *The Mesolithic in Europe*, pp. 481–491. Edinburgh: John Donald.

Thomas, J.S. (1996) *Time, Culture and Identity: An Interpretive Archaeology*. London: Routledge.

Tilley, C. (1994) *A Phenomenology of Landscape*. Oxford: Berg.

van Gennep, A. (1960) *The Rites of Passage* (original 1909). Chicago: University of Chicago Press.

White, R. (1993) 'Technological and social dimensions of "Aurignacian-Age" body ornaments across Europe.' In H. Knecht, A. Pike-Tay and R. White (eds) *Before Lascaux: The Complex Record of the Early Upper Palaeolithic*, pp. 277–299. Boca Raton: CRC Press.

Wickham-Jones, C.R. (1990) *Rhum: Mesolithic and Later Sites at Kinloch, Excavations 1984–86*. Edinburgh: Society of Antiquaries of Scotland, Monograph Series 7.

Woodburn, J. (1982) 'Social dimensions of death in four African hunting and gathering societies.' In M. Bloch and J. Parry (eds) *Death and the Regeneration of Life*, pp. 187–210. Cambridge: Cambridge University Press.

Wymer, J. (1982) *The Palaeolithic Age*. London: Croom Helm.

Zagorska, I. (1993) 'The Mesolithic in Latvia.' *Acta Archaeologica* 63: 97–117.

Zagorska, I. and Larsson, L. (1994) 'New data on the chronology of the Zvejnieki Stone Age cemetery.' *Mesolithic Miscellany* 15(2): 3–10.

Zagorskis, F. (1973) 'Das Spätmesolithikum in Lettland.' In S.K. Kozlowski (ed.) *The Mesolithic in Europe*, pp. 651–669. Warsaw: Warsaw University Press.

Zvelebil, M. (1994) 'Plant use in the Mesolithic and its role in the transition to farming.' *Proceedings of the Prehistoric Society* 60: 35–74.

CHAPTER THIRTEEN

ARCHITECTURE, IMAGINATION AND THE NEOLITHIC WORLD

RICHARD BRADLEY

INTRODUCTION: DEFINING THE NEOLITHIC

It is difficult enough to discuss the character of human creativity in the past, but it is doubly difficult to do so for the Neolithic period. That is because the very concept of 'Neolithic' is so difficult to pin down. It is a term that has often changed its meaning (Thomas 1993). Like 'culture', the word 'Neolithic' means many different things. For nineteenth-century scholars it defined a technology, a New Stone Age, distinguished from the Palaeolithic period by the use of polished stone tools. This is an approach that is no longer followed today. Second, it describes a largely new relationship between people and resources – the social changes that came about through the ownership of plants and animals. That is the sense in which the term Neolithic has been used by most prehistorians working in Europe since the early work of Gordon Childe. Thus archaeologists often write about the 'Neolithic' economy. But the term also describes a rather different material world, one in which portable objects seem to have been employed in more precisely regulated ways. Outside southern Scandinavia, it sees the first use of ceramics in Europe, and it provides convincing evidence for the long-distance movement of other kinds of artefacts, including axes. These seem to have been exchanged at a number of specialised monuments, and in many cases these objects appear to have been deposited with some formality once their use-life was over.

None of these definitions is entirely satisfactory. Polished stone artefacts are now known from the Mesolithic period. There is evidence for increasingly close relationships between humans and wild resources for some time before the adoption of agriculture, and it is becoming obvious that material culture might have been used in many kinds of social transaction from the Upper Palaeolithic period onwards. Even the creation of hoards and votive deposits seems to originate during the Mesolithic. As Julian Thomas has argued, we need to 'rethink' the Neolithic (Thomas 1991).

A further approach is to emphasise the one feature of the Neolithic world that was entirely new. In this case the term refers to a built environment for which there was no precedent in European prehistory: to the massive long houses found in some areas and to the construction of more specialised buildings in others (Whittle 1996: ch. 10). Some were residential structures and were created at the heart of the agricultural landscape, but others were more closely associated with the dead and the supernatural. What was the relationship between these different features? How were the origins of agriculture related to the changes of material culture that took place at the same time? And how close was the connection between economic changes of this kind and the impulses that led to the creation of the first prehistoric architecture in Europe?

These different definitions of the Neolithic work together in certain regions, whilst in others they fly apart. In central Europe, for example, these elements interact and there is no difficulty in discussing the character of Neolithic culture: cereal agriculture was practised in a landscape with massive domestic buildings, and these developments accompanied the spread of new kinds of material culture (ibid.: chs. 5 and 6). Around the rim of north and northwest Europe, each of these features takes on a life of its own. Material culture is employed in a highly structured manner, but it is hard to work out how this change can be related to the adoption of agriculture. The emergence of monumental architecture poses a problem too, for stone and earthwork monuments are best represented in those areas where economic change is difficult to recognise.

Domestication is both an economic relationship that connects people to resources, and a social relationship that links people with one another. Neolithic architecture creates fresh relationships too, for, whether the builders intended it or not, the erection of massive monuments would have been instrumental in creating a new sense of place and time. All buildings occupy a particular point in the terrain, and the scale of many of these structures means that they transcend the generations. Exactly the same is true of farming, which relies on a continuous input of labour directed towards a specific area of land, but the relationship between these elements was not one of cause and effect. Monuments were constructed as soon as there is any evidence for domesticates, so there is no reason to suppose that their erection was financed by the production of surplus food. If anything, the chronological relationship works the other way. Over most parts of northwest Europe, the first monuments appear some time before there are signs of agricultural expansion (Bradley 1993: ch. 1).

On the other hand, we should not separate these different elements completely. Maybe these monuments helped to associate the dead of the community with particular parts of the terrain. Perhaps the very presence of monuments in the landscape made it easier to assimilate the new sense of time and place without which farming was, quite literally, unthinkable (Bradley 1998). Perhaps, too, the use of these monuments provided a demonstration that extended beyond the individual lifespan of the continuing power of the dead and their vital importance for the living.

CREATIVE ARCHITECTURE

It is widely accepted that creative thinking brings together different elements that are not normally associated with one another: different conceptual spaces in Margaret Boden's term (Boden 1990). But it does so in a context in which the creative imagination is actually disciplined by rules. In classical music, for example, a fugue is among the strictest forms available to a composer, yet it is one that led Bach to some of his greatest achievements. Another example of the same principle is George Perec's famous novel *La Disparition*, which was written after a serious creative block. What provided the stimulus for writing it and allowed Perec to produce one of his most original works was the challenge of writing an entire book without using the letter e.

I would like to suggest that Neolithic architecture arises out of a rather similar relationship between invention and constraint. For the first time in prehistoric Europe, abstract ideas about the world were articulated in great three-dimensional constructions. Neolithic intelligence was stimulated by the physical limitations imposed by the raw materials of these monuments, among them stones of a size that had never been used before. It was in the interplay between cosmology and engineering that the creative imagination found its path.

Andrew Sherratt recently commented that the first megalithic monuments were made from components that were hardly altered from their natural state: from boulders and large rocks that were used without modification (Sherratt 1995: 245–246). He contrasts this with the deliberate shaping of building stone in other societies, but he does not develop this point. That is a pity, because in structural terms the use of such materials presents a particular kind of challenge. Megalith building is a process of 'fitting together' disparate elements selected from the natural world. The best analogy is with drystone walling. That is quite unlike other kinds of project, where the building material is shaped to conform to a pre-ordained design. One method depends on accurate planning and measurement, whilst the other relies on successful improvisation.

The Clava Cairns illustrate these points in detail. In this chapter I shall concentrate on three particular monuments: two well-preserved chambered cairns at Balnuaran of Clava, where I have excavated over the last few years, and a more ruinous circular structure that is located in between them (Figures 13.1 and 13.2). All three monuments are found towards the western limits of an extensive cemetery that originally included at least eight separate mounds or cairns. These were laid out in two

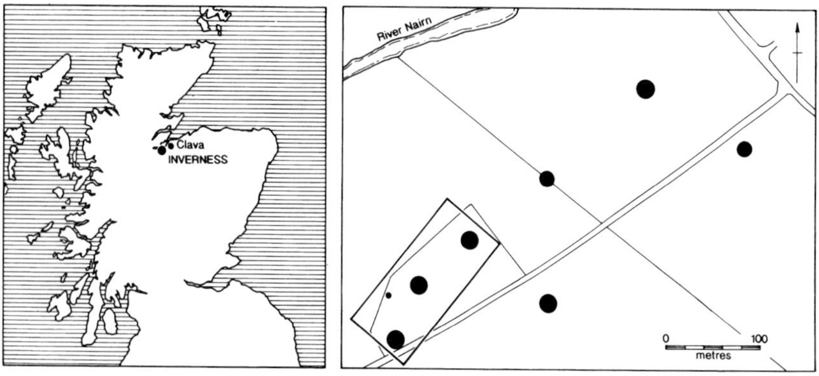

Figure 13.1 The location of Balnuaran of Clava and an outline plan of the cemetery

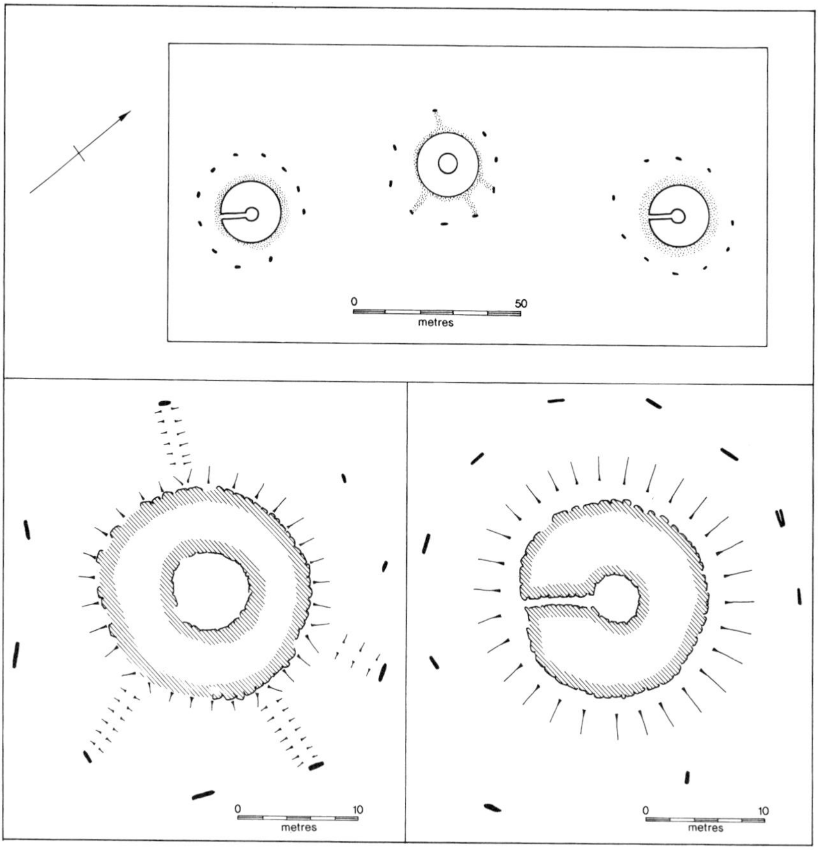

Figure 13.2 (Top) Outline plan of the three monuments discussed in the text; (left) outline plan of the ring cairn; (right) outline plan of the southwest passage grave

converging lines extending along a gravel terrace, and the pivotal point of the whole group is one of the chambered cairns. The internal chronology of this cemetery is difficult to establish and may extend from the third millennium to as late as the first millennium BC. The three cairns considered here, however, are all built in virtually the same manner and are likely to have been used over much the same period of time as one another. They seem to be among the oldest elements in the cemetery, and that is why they are so relevant to the theme of this chapter.

Having emphasised the distinctive character of Neolithic creativity, I must say something about the main elements that were brought together in these buildings: a series of raw materials with certain physical properties, and a series of abstract ideas about the relationship between the dead and the natural world. It is in the tension between the two that the Neolithic achievement had its roots.

All three excavated cairns at Clava are built out of immediately available materials, and they were constructed with an absolute minimum of effort. The final effect was all that mattered, and in seeking a means to that end the builders took what seem to have been calculated risks. They used several different kinds of stone. The most obvious are the large glacial erratics that occur in profusion within the gravel on which the site was built. They selected these for their distinctive shape and size and sometimes for their colour. The same applies to the rounded boulders that are found throughout these gravels. Again the workforce had the choice of different kinds of raw material, for these deposits include rocks that had been moved by melting ice from large areas of the Scottish Highlands. The cairns also incorporate a number of sandstone slabs that were probably quarried on the banks of the River Nairn 150 m away. We can suggest, but cannot prove, that some of this material was being reused from earlier buildings on the site.

Two kinds of monument are considered in this chapter. The first is the *passage grave* (Figure 13.3). This is a circular cairn, bounded by a stone kerb, with a corbelled chamber at its centre. That chamber was reached from outside by a low passage. Such monuments contain very small numbers of human burials. These may be either cremations or inhumations. The other type of monument is the *ring cairn* (Figure 13.4). This originally took the form of a continuous stone-walled enclosure surrounding a central space that was open to the elements. Either there was no formal entrance to the interior or, more likely, it was blocked in antiquity. On most structures of this kind, the central part of the ring cairn was filled with rubble after a number of cremations had been deposited there. Both forms of monument appear to be built on top of low platforms and both are enclosed by circles of freestanding monoliths (Figure 13.5). A characteristic of each kind of cairn is that the stones of the kerb, and those of the outer circle, seem to be graded in height. The tallest stones are generally towards the southwest and lowest are to the northeast (Figure 13.6).

The monuments at Balnuaran of Clava are the largest of this type. Here the symbolism of the two passage graves is all-important. They follow the same alignment as one another, with a third structure, a large ring cairn, in between them but offset from that axis so that this monument does not interrupt the view. Each passage grave is orientated on the midwinter sun as it sets behind a nearby hill (Figures 13.7 and 13.8). Even at this level it appears that the structure of these tombs establishes a symbolic link between the dead in the burial chambers and the annual cycle of the seasons. By linking the ancestors with the natural order in this way, the people who built the tombs placed them beyond any challenge and linked them to a sequence of events that would continue unbroken into the future.

The architecture of those two tombs emphasises this relationship on several different levels. Both cairns are enclosed within a freestanding stone circle and the passage graves are orientated towards the southwest: that is to say, they face the setting sun on the shortest day of the year. This characteristic feature is emphasised throughout the structure of both the monuments. Thus the upright stones encircling each of the cairns are graded in height, with the lowest of them to the northeast and the tallest to the southwest. Exactly the same applies to the kerbstones that retain the edges of these cairns, and at Balnuaran of Clava that principle even extends to the orthostats that form the bottom course of the burial chambers. These chambers were originally corbelled, but the drystone walling that accounts for most of their structure rests on a series of uprights. Again, they were lower towards the northeast and higher towards the southwest, where the chamber meets the passage. In each case it seems as if the structure was tilted into the setting sun.

The same point is made in other ways. The segment of the monument that was directed towards the sunset makes use of a greater proportion of red orthostats than the remaining part of each cairn. This is particularly true of the passage grave on the end of the cemetery which faces directly into the sunset, for in this case the stones were quarried for the purpose. By contrast, the backs of these two monuments, which are directed towards the rising sun, make greater use of glacial erratics. These are often of pink or white quartz which reflects the light. At the same time, each of the cairns is raised on a rubble platform, which was originally bounded by a stone circle (Figure 13.5). This gives an appearance of added height to the monument, and the same effect is enhanced by the setting of the kerbstones, which slope inwards towards the cairn at an angle of about 70 degrees. Thus the whole structure seems to taper, and again this gives the impression that the monument is higher than is actually the case. The platform was either built simultaneously with the cairns or it was constructed immediately afterwards. Certainly, it continues unbroken across the entrance to the tombs, so that the passage slopes down to the central chamber. As it does so, it becomes both higher and wider. This means that it can act like a lens, focusing the last rays of the sun on the rear of the chamber where the dead are buried.

These effects are emphasised in smaller details of the design. Looking into the monument, we have the illusion that the burial chamber is sunk into the ground, and this impression is enhanced because the lowest orthostats used in building that chamber are opposite the end of the passage. The same image is reinforced by the way in which the corbel was built, for the highest part of the chamber was not located at its centre but towards the back wall. In both the tombs, the end wall, which would have been illuminated by the midwinter sun, includes an unexpectedly high proportion of dark red slabs. This may not have been fortuitous, for there is a concentration of red boulders on the surface of the cairn at the limit of the cemetery where it faces directly into the sunset. Inside that same tomb, the rear wall includes a number of carved stones that would have been concealed from view once the corbel was built. These are found in the part of the chamber that would have been illuminated by the setting sun.

We can also consider the view looking out from that chamber. In this case, some of the visual effects are reversed. In one of the tombs, the roof of the chamber slopes down to meet the end of the passage, and the point where they meet is emphasised by a decorated orthostat. This could be viewed only by an observer positioned against the back wall of the chamber. Beyond the decorated stone, the floor of the passage rises so that the eye is directed above the surrounding area and takes in a section of the sky. The limits of the passage seem to close in towards the entrance of the tomb until the observer's attention is fixed on the point where the sun will set.

I mentioned that both the tombs were built on the same axis as one another. This means that one of the monuments was aligned directly on the midwinter sun as it sank below a nearby hill. The second cairn was directed towards the same point on the skyline, but in this case it was also aligned on the other monument. This has interesting implications, for when we calculate the original height of the chambers on both these sites, it becomes clear that another visual effect was intended. From the end cairn in the cemetery at Balnuaran of Clava, there is an uninterrupted view of the sunset; but, seen from the other passage grave, the summit of that cairn merges with the horizon. As a result, the sun appears to set on top of the tomb itself (Figures 13.9 and 13.10).

Such symbolic elements are striking, but their importance is enhanced when we realise that nearly all of them were achieved at some risk to the structure of the tombs. That is what I mean by a creative tension between cosmology and engineering. One tomb was more difficult to build because the highest part of the chamber, with its rear wall of red stone, was built on top of the lowest of the orthostats. In the same way, it was difficult to support the base of a rubble cairn when the kerbstones on one side of the monument were much lower than those on the other,

Figure 13.3

Figure 13.4

Figure 13.5

Figure 13.3 The northeast passage grave during excavation showing the kerbstones flanking the entrance, the passage, the central chamber and part of the surrounding stone circle
Figure 13.4 View of the ring cairn during excavation showing the outer kerb in the foreground, the inner kerb in the background and traces of a radial division running through the material of the cairn
Figure 13.5 The southwest passage grave under excavation, viewed from the rear of the monument. One stone of the surrounding circle is shown in the foreground and the material of the rubble platform is visible outside the edge of the cairn

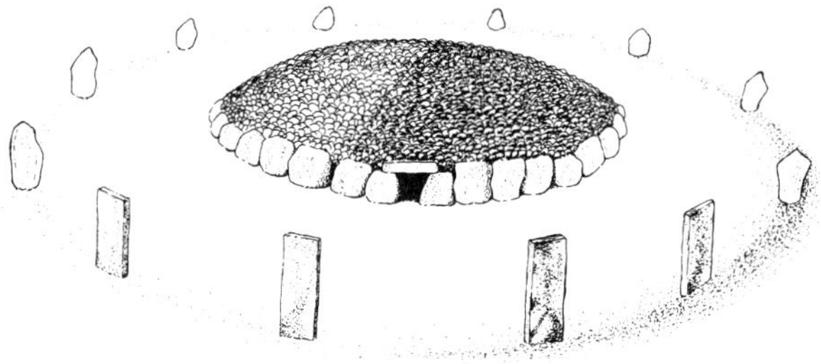

Figure 13.6(a)

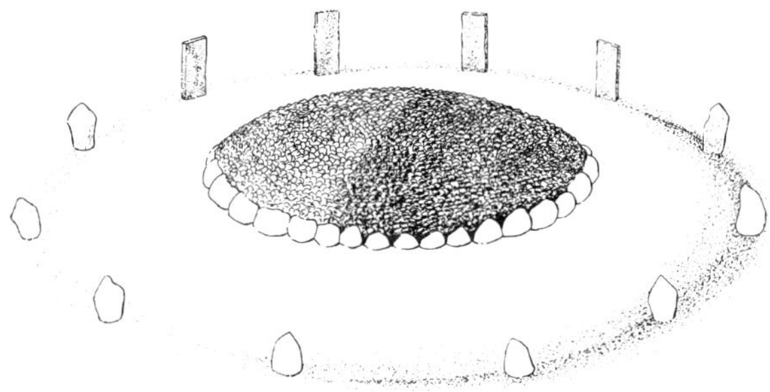

Figure 13.6(b)

Figure 13.7

Figure 13.6 The characteristic structure of a Clava passage grave (a) looking towards the entrance and (b) looking towards the rear of the monument
Figure 13.7 The midwinter sunset as viewed from the position of the southwest passage grave at Balnuaran of Clava

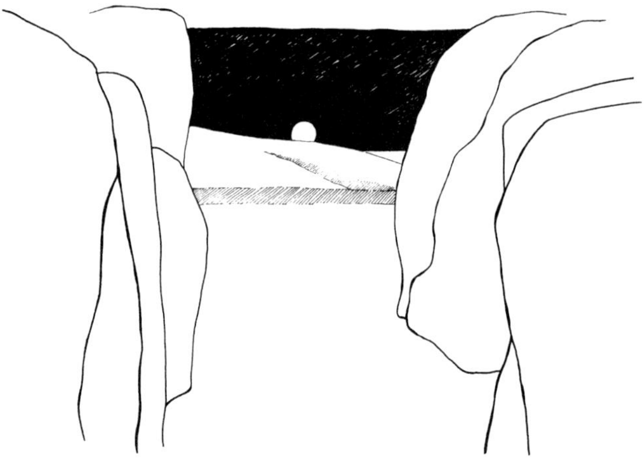

Figure 13.8

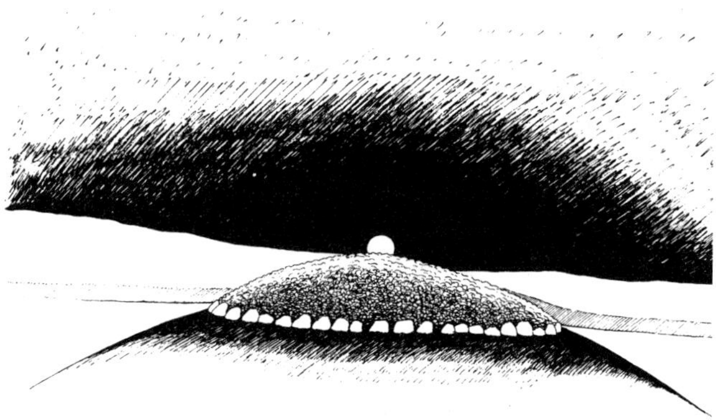

Figure 13.9

Figure 13.8 The midwinter sunset framed by the surviving stones of the southwest passage grave at Balnuaran of Clava
Figure 13.9 The midwinter sunset as viewed from the position of the northeast passage grave at Balnuaran of Clava

and the problem could be solved only by propping them between the outer material of the monument and the rubble of an external platform. This must have happened during, or very soon after, the completion of the cairn, and it was also how it was possible to create a kerb that seemed to taper inwards towards the body of the cairn. It was a risky strategy because very few of the kerbstones were provided with any sockets, and on a similar cairn at Corrymony, where the external platform was too small, the entire structure gave way (Piggott 1956: 174–184).

At the same time, each of the structural devices employed to support these monuments had a further significance. If one function of the external platform was to pin the kerbstones into place, another was

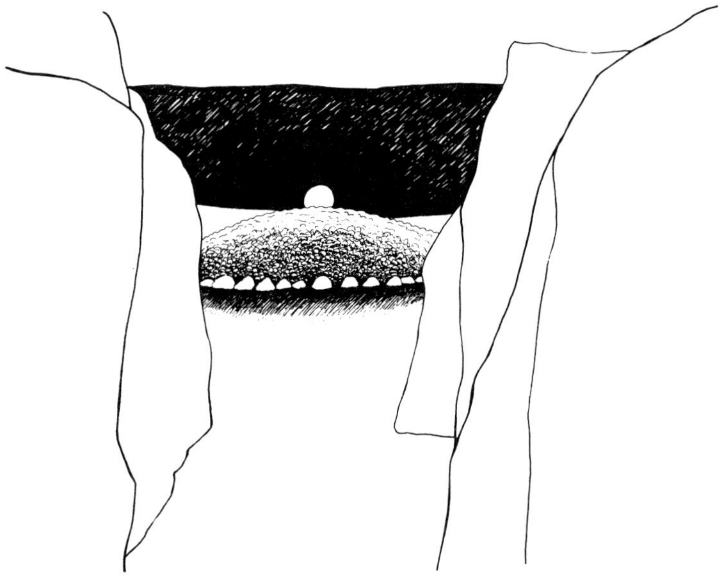

Figure 13.10 The midwinter sunset framed by the surviving stones of the northeast passage grave at Balnuaran of Clava

perhaps to raise the entrance to the tomb so that the passage became more constricted towards the entrance until it was aligned on one small segment of the horizon. Those platforms had other functions too. They seemed to raise each of the cairns above the surrounding area, and at the same time they may have provided a stage where those taking part in ceremonies around the tomb would occupy a more prominent position. The surface of at least one of these platforms seems to have been paved and both may have formed the focus for deposits of seashells and cremated bone scattered outside the kerb. They contributed to the wealth of visual illusions created by the architecture of these monuments at the same time as they held these curious structures together.

Many of the same observations apply to the central ring cairn, which is not so well preserved. This monument lies in between these chambered tombs but is offset from their axis, although two of the monoliths enclosing the cairn are in line with their entrance passages. The ring cairn is set within a slight rubble platform and at four points its outer kerb is linked to the stone circle by banks of rubble (Figure 13.2). At least one of these divisions continued as a radial division in the rubble filling of the cairn itself and linked the stones of the two kerbs (Figure 13.4). The close connection between these three circles is also reflected in the colour and shape of the stones, so that individual monoliths are paired with the closest pieces in the inner or outer kerbs.

The stones of the inner kerb consist of upright slabs that are highest to the southwest, and this effect is repeated on a larger scale in the outer kerb and the stone circle. The tallest pieces were generally quarried for the purpose and were coloured red. By contrast, towards the back of the monument more use was again made of glacial erratics whose shapes and colours reflect those of the kerbstones to the rear of the two passage graves.

At one level, the structure of the ring cairn reflects many of the concerns expressed in the building of those passage graves. Again the monument seems to rise up towards the southwest, where an arc of large red slabs faces into the midwinter sunset, and again the entire monument is associated with the remains of the dead. But in this case the final structure was a much less stable one, and its present state of decay is partly explained by structural weaknesses that were inherent in the original design.

Two of these weaknesses are particularly revealing. Both passage graves were effectively encased in massive platforms that were bounded by stone circles. At the ring cairn, the same connection was emphasised by the banks of rubble, but any platform was of very limited extent. As a result, it was not enough to support the pressure of the cairn against the kerbstones, and many of these have slipped from their original positions. At the same time, the preference for tall red stones in the southwestern sector of the kerb meant that relatively friable fragments of sandstone, apparently quarried for the purpose, were employed in this part of the monument. Some of these were poorly suited to this role and have decayed or broken up. In each case it seems as if the structural integrity of the ring cairn was compromised by the need to adhere to the symbolic system prevailing in other parts of the site.

When we consider the cemetery as a whole, it becomes evident that this interplay between symbolism and structural engineering goes even further. Every structural device at Clava had a symbolic role, and every piece of symbolism required for the correct understanding of these monuments necessitated a structural solution. The paradox of such sites is that they were built with so little effort, using material that was immediately to hand. The cairns were constructed with the minimum of work, for very few of the uprights are supported by adequate sockets and some have none at all. With so many conflicting requirements, it needed a fine calculation to produce a monument that would actually stand, yet the cairns at Balnuaran of Clava are largely intact today.

IMPLICATIONS OF THE ARGUMENT

If creativity arises out of a tension between the imagination and a set of rules, then Neolithic architecture illustrates both of these features. The Clava Cairns are by no means unusual, and similar studies of the relationship between symbolism and engineering have been conducted in other parts of Europe. There are papers that consider the sources of the raw materials used in megalithic tombs (Kinnes 1989: 40) and the likely methods of transporting them (Mohen 1980). There are numerous analyses of the astronomical significance of these monuments (Fraser 1983: ch. 15; Hårdh and Roslund 1991), and there are also studies of how they were built (Atkinson 1961). At the same time, other authors have discussed the ways in which the architecture of these buildings would have influenced the movement of the people who used these sites (Thomas 1990). Many writers have considered the symbolism of the buildings – were they copies of the houses of the living (Bradley 1996b)? – and they have also discussed the distinctive symbolism that lay behind the distribution of different raw materials in these monuments (Tilley 1991).

On one point most authorities are agreed. These structures required a considerable amount of carefully coordinated labour for their completion, and yet the design of those buildings was the result of a careful calculation that ensured that no more effort was invested in these structures than was necessary for them to stand. Sometimes those calculations went wrong and a major monument collapsed – that has been suggested at Newgrange in Ireland (O'Kelly 1982: ch. 5) – but more often the structure retained its stability over a vastly longer time than could have been conceived by its original builders. Not only were these monuments a new feature of the Neolithic landscape, they continue to have a role in the landscape today.

That is very far from treating such constructions as the consequences of an agricultural economy, for in those parts of Neolithic Europe where there is evidence of stable mixed farming, such structures rarely occur, and when they are found they do not belong to the earlier phases of settlement. On the contrary, it is precisely in those regions in which economic change seems to have been more gradual that monuments of this kind were built. They first appear at the very beginning of the Neolithic period, and for that reason they must be regarded as a central element of prehistoric material culture. Their appearance was unprecedented, and their creation may have more to tell us about the development of ancient society than the changes of food production that were taking place at the same time. This may have been the period of the 'first farmers', but it was through the union of architecture and imagination that the Neolithic world was made.

ACKNOWLEDGEMENTS

I am grateful to Margaret Mathews for preparing Figures 13.1 and 13.2 and to Aaron Watson for the reconstruction drawings.

REFERENCES

Atkinson, R. (1961) 'Neolithic engineering.' *Antiquity* 35: 292–299.

Boden, M. (1990) *The Creative Mind*. London: Weidenfeld & Nicolson.

Bradley, R. (1993) *Altering the Earth*. Edinburgh: Society of Antiquaries of Scotland.

Bradley, R. (1996a) 'Excavations at Clava.' *Current Archaeology* 148: 136–142.

Bradley, R. (1996b) 'Long houses, long mounds and Neolithic enclosures.' *Journal of Material Culture* 1: 239–256.

Bradley, R. (1998) *The Significance of Monuments*. London: Routledge.

Fraser, D. (1983) *Lands and Society in Neolithic Orkney*. Oxford: British Archaeological Reports 117.

Hårdh, B. and Roslund, C. (1991) 'Passage graves and the passage of the moon.' In K. Jennbert, L. Larsson, R. Petré, and B. Wyszomiska-Werbart (eds) *Regions and Reflections. In Honour of Märta Strömberg*, pp. 35–43. Lund: Almqvist & Wiksell.

Henshall, A. (1963) *The Chambered Tombs of Scotland, volume 1*. Edinburgh: Edinburgh University Press.

Kinnes, I. 1989. 'Megaliths in action. Some aspects of the Neolithic period in the Channel Islands.' *Archaeological Journal* 145: 13–59.

Mohen, J.-P. (1980) 'La construction des dolmens et menhirs au Néolithique.' *Dossiers d'Archéologie* 46: 58–67.

O'Kelly, M. (1982) *Newgrange. Archaeology, Art and Legend*. London: Thames & Hudson.

Piggott, S. (1956) 'Excavations in passage graves and ring cairns of the Clava group.' *Proceedings of the Society of Antiquaries of Scotland* 88: 173–207.

Sherratt, A. (1995) 'Instruments of conversion: the role of megaliths in the Mesolithic/Neolithic transition in north-west Europe.' *Oxford Journal of Archaeology* 14: 245–260.

Thomas, J. (1990) 'Monuments from the inside: the case of Irish megalithic tombs.' *World Archaeology* 22: 168–178.

Thomas, J. (1991) *Rethinking the Neolithic*. Cambridge: Cambridge: University Press.

Thomas, J. (1993) 'Discourse, totalisation and "the Neolithic".' In C. Tilley (ed.) *Interpretative Archaeology*, pp. 357–394. Oxford: Berg.

Tilley, C. (1991) 'Constructing a ritual landscape.' In K. Jennbert, L. Larsson, R. Petré and B. Wyszomiska-Werbart (eds) *Regions and Reflections. In Honour of Märta Strömberg*, pp. 67–79. Lund: Almqvist & Wiksell.

Whittle, A. (1996) *Europe in the Neolithic*. Cambridge: Cambridge University Press.

CHAPTER FOURTEEN

The conditions of creativity for prehistoric Maltese art

CAROLINE MALONE AND SIMON STODDART

The Maltese islands (Figures 14.1 and 14.2), located some 85 km south of the island of Sicily in the central Mediterranean, are renowned amongst archaeologists for one of the most marked, distinctive and original developments of artistic creativity in prehistory. During the mid to late fourth and early third millennia BC, a period of consolidation nearly two millennia after the transition to agriculture, there was the long-term maintenance of a particular range of artistic and architectural styles that differed completely from those of neighbouring contemporary societies. The large and fertile neighbour, Sicily, to the north, had no such evident artistic creativity during the same period. In Malta, architecture became monumental in form, employing interconnected lobed apses on an axial frame, forming what are generally accepted as temple structures. Human representation was marked by a characteristic corpulence of the human form. Natural representation included abstract vegetal forms of an almost modern quality. All these elements added up to a highly distinctive repertoire of artistic forms.

This chapter aims to investigate the conditions for the creativity behind these developments of elaborate prehistoric art which were persistent for more than a millennium, but had no long-term succession in Malta or beyond. The conditions were very specific to a specific historical context in the development of the Maltese islands and have not been repeated in later times. The conditions existed at three levels: the relationship to the rest of the Mediterranean, the social and ideological conditions of Maltese society, and the degree of individuality allowed in some representations of art.

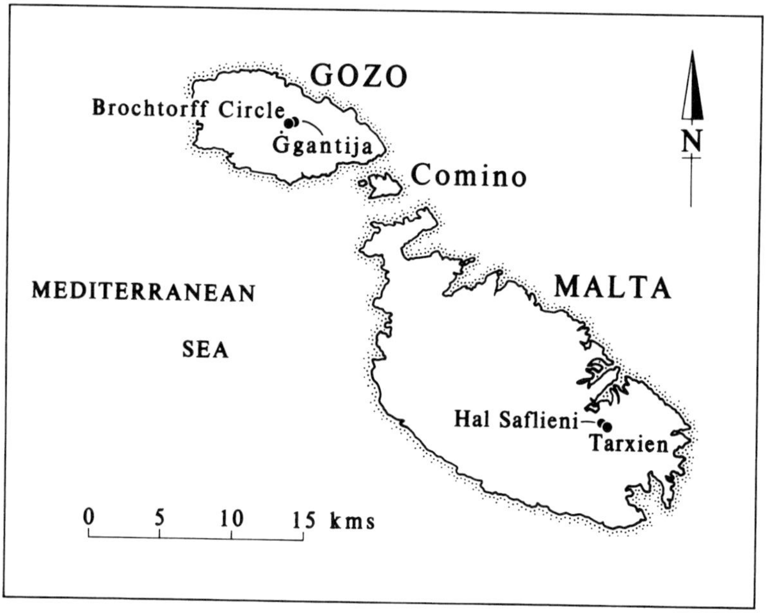

Figure 14.1 Map of the Maltese islands showing major sites discussed in the text

Source: Drawn by Steven Ashley

Figure 14.2 The Xaghra plateau in Gozo from the air

MEASURING THE VALUE OF CREATIVITY

In spite of the intrinsic interest of prehistoric Maltese art, other artistically creative phases of prehistory are much better known outside the narrow world of archaeological specialists. Discussions on prehistoric art in Europe are considered in an implicit hierarchy of significance which can be measured in the relative number of pages devoted to them. A standard work on *Prehistoric Art in Europe* (Sandars 1985), concentrates primarily on the Celtic art of the Iron Age (more than 25 per cent), an issue of current perceived cultural significance to modern Europeans, and the symbolic explosion of the Upper Palaeolithic (slightly less than 25 per cent), a threshold in the creative capacity of modern humans. The Maltese islands in prehistory receive less than 1 per cent of these same pages.

Why should this injustice be perpetrated? The tradition of study of western art is one of grand cumulative development, as revealed in the great Christian image of the stained glass of the southern transept of Chartres: the evangelists of the New Testament rest on the shoulders of the prophets of the Old Testament. If art lies outside the tradition of cumulative development, it has no role in the grand scheme. Prehistoric Maltese art had no place in grand schemes of European art, and attempts to bring it into such a scheme are generally discredited (Gimbutas 1982, 1993). Prehistoric Maltese art may have developed as its own distinctive tradition over more than 1,000 years, but it had no local or international legacy. The lack of influence on other traditions should not in our opinion be a measure of the level of creativity. The conditions for creativity have simply never been repeated, in Malta or beyond.

One positive element of this undervaluing of prehistoric Maltese art is that it has never formed part of the trade in ancient art, with the consequent dispersal of information. The notorious Ortiz collection contains no Maltese art, but does include art broadly similar in concept (Ortiz 1994: cat. no. 43–44), collected only because it forms geographically (but not culturally) part of that grand scheme of alleged European continuity located in Greece.[1] Unlike Ortiz, we do not need to look beyond Europe (in search of ethnography) to explore an exciting diversity of creative conditions; furthermore, to reconstruct those conditions, we are dependent on information on provenance and context that the very act of assembly of the Ortiz collection has destroyed for the objects it contains. Collections like that of Ortiz are vandalistic towards the discovery of the motivating creativity behind the art they contain.

MEDITERRANEAN CONDITIONS

The long-term prehistoric (and subsequent) trajectory of the Maltese islands is that they were placed in regular contact with the surrounding Mediterranean, particularly with the non-African rim (especially with

Sicily between 5500–3800 BC and 2500–800 BC, and with the Levant between 800 BC and the arrival of Rome) (Stoddart, in press). Agriculture was a prerequisite for the colonisation of the relatively small islands of Malta (310 km^2), and cultural connections were maintained with Sicily, the source of the colonists, for more than 1,000 years (Tusa 1992: 202). The Early Neolithic Stentinello style pottery of Sicily has strong affinities with the Ghar Dalam style of Malta, although less marked similarities persisted throughout much of the later phases of the Neolithic. Clear cultural connections were re-established with the onset of the Bronze Age in the central Mediterranean area, following *c.*2500 BC (ibid.: 343). Some of the characteristic shared features are bossed bone plaques (probably knife handles for bronze blades), increased use of metalwork and the exchange of pottery (Procelli 1996: 93). At the end of the Bronze Age, the Maltese islands were taken over by the Phoenicians in a strategy of complete incorporation that was facilitated by the relatively small size of the islands.

The one potential exception to the overriding pattern of interconnectedness is that of the temple building period (3800–2500 BC), which is characterised by the distinctive cultural and artistic achievements that are the subject of this chapter. Even in the case of this potential exception, many scholars emphasise the constant linkage of Malta to the rest of the Mediterranean either as an intense network or as the residual remains of a pre-existing network, as we have recently outlined (Malone *et al.* 1995: 2–4) and which are presented in terms of the consequence for interpretations of creativity below.

Although we believe that Malta was culturally isolated during this period, its cultural developments are paralleled elsewhere in the Mediterranean Neolithic world with regard to economic consolidation, expansion and marked social change. From a thinly populated world of mobile hunter-gatherers in the period 7000–6000 BC, the massive and virtually irreversible changes wrought by sedentary food-producing societies in the region between 6000 and 4000 BC triggered a series of technological and social transformations in a complex mosaic of different rates of change in different zones of the Mediterranean. These changes have been well rehearsed elsewhere (e.g. Lewthwaite 1986), and we can simply note that they include the development of village life and the close social proximity of larger communities. Life became quite crowded in comparison to earlier times, and various activities and social conventions developed to cope with the stresses and complexities of the annual farming cycle. There was little escape from the annual routine or from neighbours.

Much contemporary anthropological literature describes the tensions and pettiness of life in the enclosed environment of small-scale villages (e.g. Boissevain 1965). Frequently, religious specialists and shamans emerge to manipulate and mediate between both rival human groups and the capricious natural world and its spirits. Mediation between these forces that controlled the fertility of both agriculture and the human community required tangible apparatus and specialist paraphernalia that embodied the beliefs and fears of the community. It is in this milieu of farming-based, small-scale society that the so-called 'mother goddess' phenomenon emerged in prehistoric Europe and the Near East. The chronological sequence of development from *c*.7000–2500 BC is long, but there is a consistent pattern between the transition to agricultural settled life, developing social complexity and the use of figurative art, especially modelled anthropomorphic figurines.

The scholars who emphasise the interconnectedness of Malta are mainly early in date. Fergusson (1872: 28) pieced together the available evidence of his time into a stylistic pattern of megalithic architecture across the Mediterranean and the rest of Europe as 'one continuous group, extending in an unbroken series, from the earliest to the latest'. The Maltese islands were seen as an adjunct constructed with metal tools, but equally strongly 'mixed up in all the works treating of the subject with Druidical remains' (ibid.: 415ff.). The motives of creativity were not clearly indicated, but he stresses funereal rituals, noted their distinctiveness and originality and placed their date at the time of the Trojan Wars (ibid.: 426–427). In his diffusionistic framework of expanding architectural styles, Fergusson had to find a source of influence, and since he found the creativity difficult to parallel, he placed the source in the unresearched zone of (north) Africa: 'They are too unlike anything else in Europe' (ibid.: 426).

Zammit (1930: 122), the great innovator of archaeology in the Maltese islands, envisaged Malta as 'the Holy island of neolithic faith, the half-way house of the early mariners, who trusted themselves to their frail wooden craft, full of hope in a protecting power'. The creativity of the Maltese art was based on the devotion and faith of pilgrimage and the participants implicitly displayed their faith in an imagery completely different from and foreign to that of their homelands.

Maringer (1956: 152–159) emphasised the dourness and strictness of a chthonian cult that was later revitalised by a Great Mother religion from the Near East. The end-result was thus 'the product of a mingling of western and eastern influences'. The creativity behind the art and architecture was clearly 'the apparent power of their religious impulse'.

Bernabò Brea (1960) stressed the role of Malta as an 'obligatory port of call', based, no doubt, on the later, but intermittent, historical parallels. The 'flowering of Maltese culture' was explicable only 'as the result of an intensive maritime and commercial activity'. The poor resource status of Malta required the Maltese to develop external contacts. The implication is that the creativity is partly external, but in turn Malta, with memories of the grand Malta-centric schemes of Ugolini (1934), was 'at the root of all megalithic architecture of western Europe'.

More recently, McConnell (1985) has presented what might be characterised as the Medici model of Malta. The patrons of the Maltese temples derived the means for their temple investments from the profits of textile production using raw material from Sicily. This idea glosses over problems of bulk transport and lack of evidence for weaving production on the Maltese islands. It is a model for the time of the Knights of St John, the second great period of architectural and artistic creativity on the Maltese islands, not for the fourth and third millennia BC. This historical period of Christian creativity should not be allowed to impinge too deeply on the reconstruction of prehistoric creativity. The conditions of creativity in the Maltese islands are unique and specific and not part of the grand schemes of cumulative development that dominate the history of Western art.

Gimbutas (1982, 1991, 1993) considers Maltese art and architecture to be the late and residual product of a pre-existing and pervasive pattern of matriarchal ideology that extended over much of Old Europe. She envisaged an enduring view of the world connected with cycles of nature and the human body, which persists throughout space and time with little variation. Statements such as '[the temples are] actually expressions of the regenerating body of the Goddess with enormous egg-shaped buttocks' (1991: 262) provide deceptively simple explanations of the creativity of the Maltese islands. Multiple assumptions are made about the female attribution of the images and their linkage to both matriarchy and a wider social and cultural world, and can easily be criticised (Malone *et al.* 1995). The 'mother goddess' interpretations play down the degree of local creativity in the Maltese islands, both at the level of society (since inspiration is from Old European civilisation) and at the level of the individual (since the matrifocal inspiration is all-pervading). Few scholars hold these views seriously and most are quite explicit in their outright rejection: 'reweaving a fictional past with claims of scientific proofs is simply irresponsible' (Meskell 1995, 74); 'this is quite frankly an idea picked up from the dustbin of longtime discarded anthropological theories' (Haarland and Haarland 1995: 112).

More recently stress has been laid on the independence and isolation of the Maltese islands during this flourishing period of creativity. Evans (1959: 158) was amongst the first to stress the apparent isolation, but warned against interpretations of extreme isolation: 'a kind of lotus-eating community, turned in on itself and wrapped up in its bizarre cults . . . but this is only one side of the picture'. Renfrew (1972: 1973) pressed home the consequences of calibrated radiocarbon chronologies for the Maltese islands and emphasised the potentially independent origins of Maltese creativity. Similarities in material culture to Sardinia, stressed by some (Bray 1963), were now considered parallel developments. Implied competition between chiefly groups, with their capacity to mobilise manpower, engage in group solidarity and organise redistribution, led to the increasing sophistication and elaboration of the artistic forms. It is one conclusion to emphasise independence; it is another to stress relative isolation, as we have sustained in recent work (Stoddart *et al.* 1993). The strong cultural identity of prehistoric Malta is clear, but the wider theoretical debate about the permeability of boundaries in prehistory remains open, despite having been running for nearly 20 years (Hodder 1979). Do boundaries of material culture accompany isolation, or instead the contrary conditions of interaction? No definitive rules can be established and the precise local conditions must be addressed in each case. In the case of Malta, after analysis of the local conditions and the available current evidence, we have concluded that the situation comprised one of relative isolation in concurrence with strong boundaries formed by material culture (Stoddart *et al.* 1993). The import and deposition of greenstones, obsidian and other raw materials appears to have become increasingly rare. The same conditions appear to have been widespread in the central Mediterranean during the fourth and early third millennia BC (Malone 1986), but such conclusions require further testing from securely dated deposits where the effects of the residual products of trade can be excluded. These conditions of isolation do, however, form the basis for our interpretations of creativity.

One author stresses the indigenous contribution even more strongly for this period, but in a different manner. Bonanno (1996: 53) has emphasised that a scholar who has 'lived all his life in the Maltese social and physical environment, which is in many ways similar to that obtaining in prehistory, is much better equipped to empathise with, and therefore to make a more faithful judgement of a society'. In other words, for the purposes of his article, prehistoric Maltese creativity can be understood only by the modern Maltese citizen. However, in a previous article, he supports the case for social discontinuity, stressing that the nature of the Maltese physical environment led to discontinuity and disruption in its historic development and presumably to radically different types of creativity after each new invasion following social and environmental collapse (Bonanno 1993).

CYCLES OF DIMENSIONS OF ARTISTIC INVESTMENT AND CHANGING SOCIETY AND IDEOLOGY IN MALTA

Between 3800 and 2500 BC, a period of some 1,300 years, the material culture of the Maltese islands underwent a profound transformation in three principal dimensions: the degree of Mediterranean procurement of raw materials and symbolic linkage, the mobilisation of manpower and the detail of craftsmanship (Figure 14.3).

In the Zebbug period (*c.*4100–3800 BC), the main evidence is funerary. Investment in architecture was restricted to small rock-cut tombs. There was no mobilisation of manpower on the scale that was to be seen in succeeding periods. Craftsmanship comprised the carving of schematic

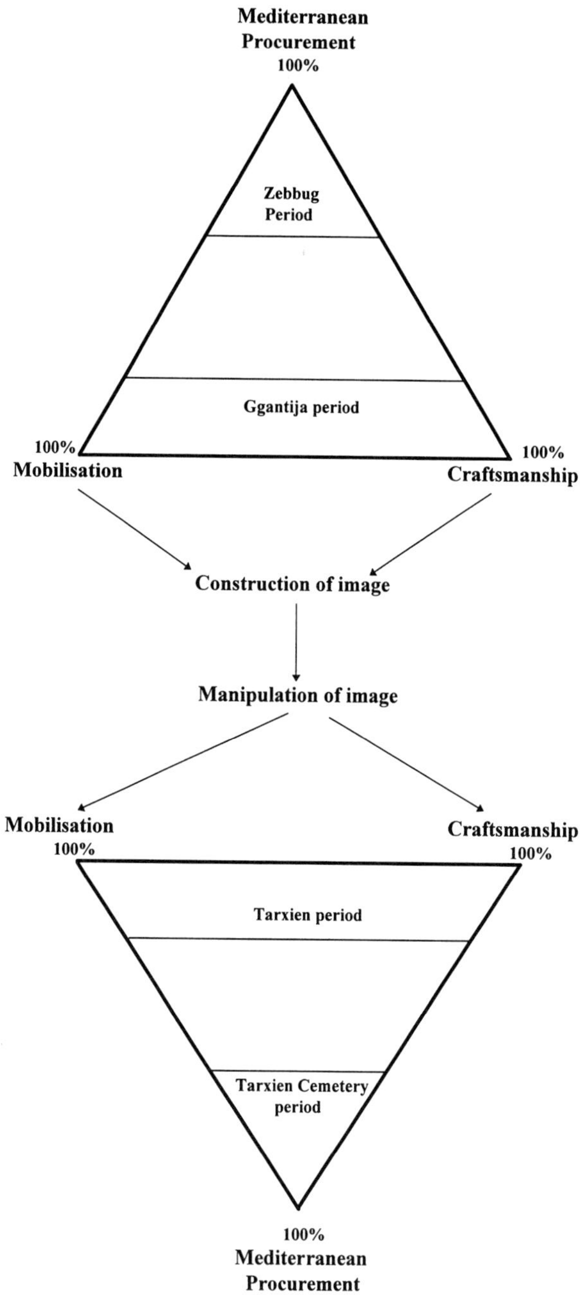

Figure 14.3 A model of cyclic creativity from the Zebbug period until the Tarxien Cemetery period

human heads in stone (Zebbug tomb, Brochtorff Circle at Xaghra tomb), small bone pendants (Brochtorff Circle at Xaghra tomb) and stick figures on pottery (Zebbug tomb and Brochtorff Circle at Xaghra tomb). There was little refinement in the construction of the human image. Only the production of stone axes demonstrates some considerable craftsmanship, and these axes, or at least their constituent materials, were procured from outside the Maltese islands from the igneous rock in mountains in Sicily, Calabria and even beyond. We have suggested that there was competition for the procurement of these raw materials and that the family or lineage success of this competition was demonstrated in funerary rituals (Stoddart *et al.* 1993: 7). We suggest that families rivalled each other in the procurement of raw materials and we can indirectly measure their relative success by looking at what they placed in tombs. When they failed to gain access to sufficient raw materials, they made faithful copies of axes in local limestone which, although functionally useless (lacking sufficient cutting edge), readily fulfilled the ritual requirement of providing substitute items of symbolic value for the dead. In this period, the stylistic attributes of the material culture of the Maltese islands still fitted easily within a wider Mediterranean ambit.

In the Ggantija period (*c.*3600–3000 BC), a new creative image was constructed, we contend, under conditions of relative isolation from the rest of the Mediterranean. This image owed nothing to the rest of the Mediterranean either in terms of the procurement of raw materials or in terms of stylistic attributes and ideology. A major focus of this image was what, borrowing from classical imagery, has been designated the temple (Figures 14.4 and 14.5). This structure typically had a dominating facade constructed from closely fitting megalithic blocks. The entrance was placed in the centre of the facade, giving a rising line of sight focused on an altar. Apses were typically placed at right angles to the main line of sight, providing an opportunity for restricted visibility from the entrance. The nature of the roofing of these structures is problematic, but it is unlikely, on engineering grounds, that the larger spaces could have been entirely roofed in stone. The secondary focus of this image was the sculptural art that filled these temples. The component of human representation can be variously categorised. One simple division is into large fixed figurines, self-standing and hand-held portable curated (re-usable) figures and portable diposable figures (Figure 14.6) (Malone *et al.* 1995: 6). To the representation of the human form, one must add animal and vegetal representations.

We suggest that early rivalry in procurement of raw materials was redirected towards the competition between factions in the construction of rival temple structures within the same community. A central community focus was maintained in the main burial monument, but internal rivalry tolerated in temple construction. This is an approach that draws strongly on analogies provided by the creativity of the modern Maltese *festa*

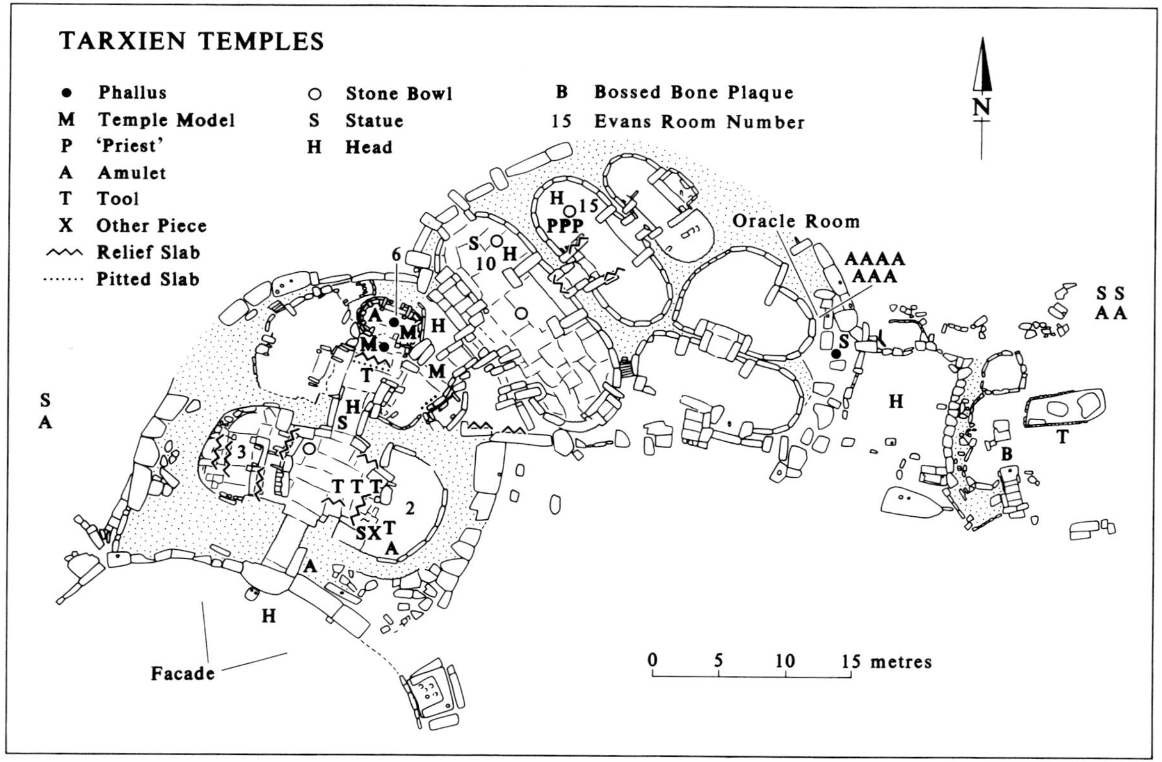

Figure 14.4 The temples of Tarxien
Source: Drawn by Steven Ashley

Figure 14.5 The Ggantija temples from the air

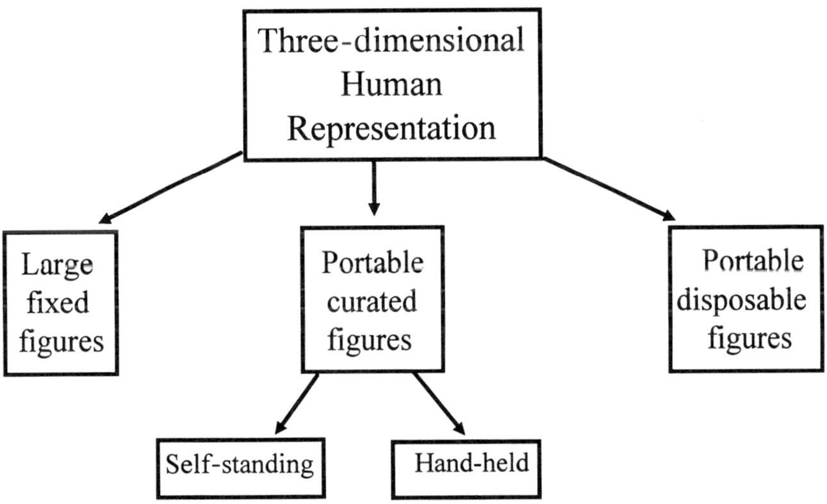

Figure 14.6 A classification of human representational sculpture from the Maltese islands
Source: After Malone *et al.* 1995

(Boissevain 1965). In the conditions of dense social networks provided by small island communities (arguably unaffected by modern urban conditions), both inter-community and intra-community rivalry existed over the production of the most elaborate fire-works and other visible elements of creative performance. In the modern period, as in the prehistoric period, the creative motivation was religious, but set within particular social conditions. In the Ggantija period, exotic materials were still in use, but it is suggested that these were mainly residual from previous exchange activity. Creativity was directed towards the modelling of local materials (limestone and clay) into new elaborate architectural and figurative forms. Our interpretation of recent work by Skeates (1995) reinforces the idea that supplies of exotic raw materials were limited in the Ggantija and succeeding Tarxien periods. Previously accumulated supplies of greenstone were reworked by a parallel process of whittling down and sacralisation (a transformation from a strongly functional to a sacred use). Axes once of a functional size were transformed into small perforated pendants concentrated in private ritual contexts (e.g. cache in the deeper recesses of Tarxien temple) (Evans 1971). The creative force was sacralisation, a product of the religious fervour that has long been identified in the Maltese islands.

Whatever one's interpretation of the relative isolation of Malta in this period, one conclusion is clear: the new artistic image of Ggantija was profoundly Maltese and dependent on the particular local social conditions for the motivation of its creativity. Religion was the creative force intensified, we suggest, by competing factions within prehistoric Maltese society (Bonanno *et al.* 1990). In this interpretation, we are again following closely the interpretations of dense personal networks by social anthropologists (Boissevain 1974). Below a demographic threshold of 3–4,000 for a corporate group, networks of relationships were likely to be intensely personal, and individuals would have been likely to have recruited personal factions. We interpret that competition between these factions is visible in the groupings of temples and associated art. The factional rivalry was contained within certain religious guidelines. The religion focused on the ideal of corpulence and other signs of natural vitality (phalli, animals and vegetal motifs). Art reinforces the special status of religion and should be conceived of as but part of a wider performance (Dissanayake 1995) that does not always leave material traces. Maltese religion was strongly connected to ritual largesse where decorated, carinated, single-handled ceramic bowls of a standard form were employed to make offerings, and give some indication of the wider arena of performance and creative activity.

In the Tarxien period (3000–2500 BC), it appears that this established image of art was extensively manipulated by a more restricted group of society. Some factions were more successful in their political organisation than others. The details of timing are difficult to establish, since most excavations of the monuments took place before stratigraphic excavation, and thus no rigorous relative dating of the monuments had been achieved. However, it is clear that although the general axial structure of a Maltese temple remained constant, access to space changed. It appears that the monuments became more restricted in access through time, with the addition of barriers to lines of sight and the creation of foci of activity separate from the wider open spaces in front of the facades of the monuments.

The single burial focus that had already been in existence in the Ggantija phase was now probably paralleled by a prominent single temple monument. It is at this stage that incipient hierarchy can be envisaged, although no social differentiation based on funerary evidence has been detected.

CONFORMISM IN MALTESE ART?

Most studies of Maltese art (e.g. Ridley 1976) have focused on those components that correspond to the public face of society and (particularly in Tarxien times) demonstrate vertical rhetoric, i.e. those that reflect hierarchical distinctions in society. These studies suggest a considerable conformism in Maltese art, a lack of creative variation, determined by religious constraints. A check-list of particular elements can be constructed that characterises this facet of Maltese art. Some caution must be applied, however, since although the sample of some characteristics is fairly large (e.g. temple structures), the sample of sculptural forms can be more restricted in some classes.

The architecture is highly recognisable. The shorthand terminology of temple is not inappropriate, given the basic canons of convention. A basic pattern was combined to form multiples and elaborations of the original design, best represented by some of the simple temple models constructed by the builders themselves (e.g. Mgarr [Evans 1971: 35, Mg/S.1; Pace 1996: 68, A1]). The concave facade was pierced by a central opening that led along a main axis communicating towards laterally arranged apses. Construction was of massive stone blocks. Allowing for variations based on the local transport distance, coralline limestone was used for the bulk of construction and the more easily worked globigerina limestone for architectural and artistic detail.

Amongst the figurative art, corpulence is a major feature (see some of the examples in Figure 14.7). This is represented most prominently in the lower parts of the body, especially in the upper thighs, the buttocks and the upper arms. The distribution of fat is characteristic: abundance generally contrasts with absence in the breast area and in the lower legs (e.g. Hagar Qim [Evans 1971: 91, Q/S.13–18]). Heads are frequently transposable, that is, formed of a separately constructed stone block that can be matched with another body and moved. These heads tend to have short cropped hair gathered in a pig-tail (e.g. Tarxien [ibid.: 144, T/P.1006]), Ggantija [ibid.: 184, G/S.2]) and Hal Saflieni [ibid.: 62, S/S.38–9]). When dressed, statues frequently have (long) pleated skirts that cover the lower part of the body (e.g. Tarxien [ibid.: 120, 144, T/P.1006]).

More abstract art also has distinctive forms. Spirals are prominent, although these do range from the highly stylised (e.g. Bugibba [ibid.: 112, Bu/S.1]) to the much more elaborate forms found at Tarxien (ibid.: plates 16–21), which may have their inspiration in plants and even animal

Figure 14.7 Range of artistic creativity from the Brochtorff Circle (not to scale)

Source: After Steven Ashley

horns. At a further level of abstraction, pock-marking distributed across stone slabs is also very frequent.

Some conformism is also shown in the bowls employed for ritual largesse. As Evans puts it, in describing the characteristics of the Tarxien ceramic repertoire:

> The most common shape by far is that of the carinated bowls, generally with triangular handle, which descend from the Ggantija phase . . . the majority seem to have been medium sized bowls. The numbers of fragments found

> show that they were made in great quantities, and suggest that they may perhaps have been used in the temples as containers for food offerings, and perhaps ceremonially broken.
>
> Evans (1971: 220)

These are canons determined by visibility. In pre-literate societies, art can replace literacy as a medium of communication. However, its power is limited or expanded by its range of receptivity, namely the field of sight. This is recognised by the fact that much early literacy has a strongly visual quality in an attempt to provide mutual reinforcement before the full flexibility of writing was exploited (Marcus 1992). Outside this field of public receptivity, local creativity of art can be allowed to flourish to a greater extent. The difference is relatively subtle in the case of Maltese art, since vertical hierarchy was probably never starkly developed. It is, therefore, difficult to find strong differentiation.

LOCAL INDIVIDUALITY

The role of individual creativity in Maltese art was somewhat increased outside the major public arenas. The temple provides the most public arena: the burial hypogeum the most private. One classical archaeologist has attempted to discern the individual in Maltese art (Bonanno 1996), but this type of approach lays insufficient emphasis on the relaxation/imposition of social constraints according to the context of creation and display of art.

Many of the smaller art objects have been considered curiosities. However, these curiosities make much greater sense in the context of their frequent inclusion in burial deposits as individualised offerings with deceased individuals who, as part of the burial ritual, lost their individuality as part of the reworking of burial deposits as new collective packages. The precise context of some of these small portable and disposable objects has become clear in the excavation of the Brochtorff Circle at Xaghra by the Anglo-Maltese team between 1987 and 1995. Certain parts of the burial hypogeum, particularly a central deposit of partly disarticulated remains, contained distinctive art objects: small animal phalanges with carved human heads, a striking clay head with tightly restrained hair, painted black, and white painted *tears* and a strange terracotta snail with an attached human head (Figure 14.8) (Pace 1996: 81, K15). A pair of stylised clay female(?) torso pendants came from an inner part of the hypogeum (Figure 14.9) (ibid.: 81, 12). A reconsideration of the Hal Saflieni hypogeum artefacts reveals a whole suite of smaller *objets d'art* (in spite of less systematic recovery and little knowledge of the precise context). These include bird pendants (ibid.: 73, F3), oxen figurine pendants (ibid.: 74, F10) and a clay lizard (ibid.: 74, F13).

Another level of individuality is the frequent representation of small-scale human heads. There may be some selection by early excavators of

Figure 14.8 Snail with anthropomorphic head from the Brochtorff Circle

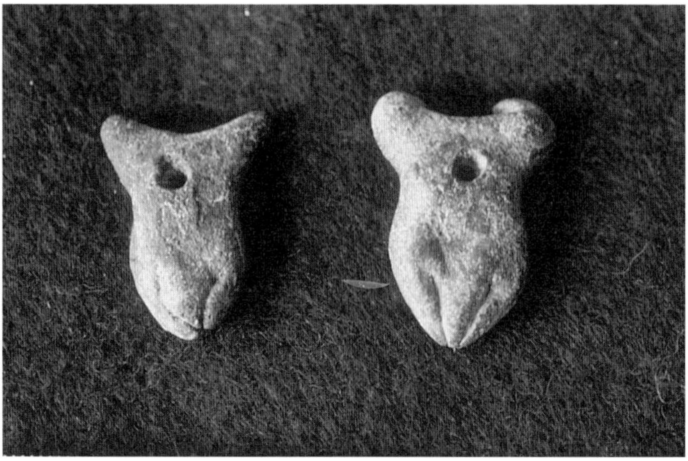

Figure 14.9 Female torsos(?) from the Brochtorff Circle

broken heads, but their high frequency even in more recent excavations suggests that this part of the body was deliberately placed in both temples and burial places. As is so often the case in Malta, there was a strong linkage in the use of material culture between temple and burial place, but their variation of detail appears to suggest some representation of human diversity on death. It is the face that provides the individual human identity and it was in the construction of the image of the face in various small-scale media that some relaxation of some of the canons of the large-scale sculpture took place. Hair, symbolic of regeneration and given particular treatment on death in many societies, was still cut short and held together tightly on the head, but facial features were represented with much greater freedom. Representations in clay have especially facilitated this variation in representation of individual identity. A collection of small heads (maximum dimensions from 2.3 to 8.5 cm) from Tarxien shows considerable difference in the treatment of the eyes, nose and chin (ibid.: 85–86, U3-U12).

The recent cache discovered at the Brochtorff Circle provides a major contrast to this. The cache of nine objects contained six plaque figurines, in various stages of manufacture. The three most finished examples had most elaborate detail in the face, but this detail involved the construction of stark and simple lines that had a generic rather than individual quality. These objects were probably involved in the funerary rituals but transferred between individuals as they became part of the funerary ceremonies. The individual was represented, but at a symbolic, abstract level.

CONCLUSION

We sustain that the motivation behind the creativity of Maltese art can be assessed at three different levels. Mediterranean conditions in the fourth and early third millennia BC led to a pattern of isolation of the Maltese islands that was exceptional in the rest of their prehistoric and historical development. Under these conditions, the introspective local communities fostered social and ideological conditions that favoured a flowering of creativity. This creativity was, however, constrained by canons of conformity that were released only in more private and, particularly but not exclusively, funerary domains. Small scale sculpture focused on either the head or persona of the deceased, and, more rarely, imaginative selections from the natural world showed relaxation of the public canons. Creativity thus has a multi-dimensional quality: the overall conditions and the level of tolerated variation from a cultural norm.

NOTE

1 The provenance is also alleged, but the reasons are rather different.

REFERENCES

Bernabò Brea, L. (1960) 'Malta and the Mediterranean.' *Antiquity* 34: 132–137.

Boissevain, J. (1965) *Saints and Fireworks: Religion and Politics in Rural Malta.* London: Athlone Press.

Boissevain, J. (1974) *Friends of Friends. Networks, Manipulators and Coalitions.* Oxford: Basil Blackwell.

Bonanno, A. (1993) 'Tarxien and Tarxien Cemetery. Break or continuity between temple period and Bronze Age in Malta?' *Mediterraneo. Revistade Estudos Pluridisciplinares sobre ad Sociedades Mediterranicas* 2: 35–47.

Bonanno, A. (1996) 'The artist in prehistoric religious ritual: servant or master?' In A. Pace (ed.) *Maltese Prehistoric Art. 5000–2500 BC*, pp. 53–58. Valletta: Patrimonju Publishing.

Bonanno, A., Gouder, T., Malone, C. and Stoddart, S. (1990) 'Monuments in an island society: the Maltese context.' *World Archaeology* 22 (2): 190–205.

Bray, W. (1963) 'The Ozieri culture of Sardinia.' *Rivista di Scienze Preistoriche* 18: 155–187.

Dissanayake, E. (1995) *Homo Aestheticus. Where Art Comes From and Why*. Seattle: University of Washington Press.

Evans, J. (1959) *Malta*. London: Thames & Hudson.

Evans, J. (1971) *The Prehistoric Antiquities of The Maltese Islands: A Survey*. London: Athlone Press.

Fergusson, J. (1872) *Rude Stone Monuments in all Countries: Their Age and Uses*. London: John Murray.

Gimbutas, M. (1982) *The Goddesses and Gods of Old Europe*. Berkeley: University of California Press.

Gimbutas, M. (1991) *The Civilisation of the Goddess: The World of Old Europe*. New York: HarperCollins.

Gimbutas, M. (1993) 'Images of old European religion.' In G. Burenhult (ed.) *The Illustrated History of Humankind. American Museum of Natural History. Vol. 2. People of the Stone Age. Hunter-Gatherers and Early Farmers. Old World Civilisations*, pp. 84–85. New York: HarperCollins.

Haarland, G. and Haarland, R. (1995) 'Who speaks the Goddess's language? Imagination and method in archaeological research.' *Norwegian Archaeological Review* 28 (2): 105–121.

Hodder, I. (1979) 'Economic and social stress and material culture patterning.' *American Antiquity* 44: 446–454.

Lewthwaite, J. (1986) 'The transition to food production: a Mediterranean perspective.' In M. Zvelebil (ed.) *Hunters in Transition. Mesolithic Societies of Temperate Eurasia and Their Transition to Farming*, pp. 53–66. Cambridge: Cambridge University Press.

McConnell, B.E. (1985) *San Cono-Piano Notaro-Grotta Zubbia Ceramics in Sicilian Prehistory*. Unpublished PhD dissertation, Brown University. Ann Arbor: University Microfilms International.

Malone, C.A.T. (1986) *Exchange Systems and Style in the Central Mediterranean, 4500–1700 BC*. University of Cambridge, Unpublished PhD dissertation.

Malone, C., Stoddart, S. and Townsend, A. (1995) 'The landscape of the island goddess? A Maltese perspective of the central Mediterranean.' *Caeculus* (Papers on Mediterranean Archaeology, Archaeological Institute, Groningen University) 2: 1–15.

Marcus, J. (1992) *Mesoamerican Writing Systems. Propaganda, Myth and History in Four Ancient Civilisations*. Princeton, NJ: Princeton University Press.

Maringer, J. (1956) *The Gods of Prehistoric Man*. London: Weidenfeld & Nicolson.

Meskell, L. (1995) 'Goddesses, Gimbutas and "New Age" archaeology.' *Antiquity* 69: 74–86.

Ortiz, G. (1994) *In Pursuit of the Absolute Art of the Ancient World from the George Ortiz Collection*. London: Royal Academy of Arts.

Pace, A. (ed.) (1996) *Maltese Prehistoric Art. 5000–2500 BC*. Valletta: Patrimonju Publishing.

Procelli, E. (1996) 'Sicily between the early and middle Bronze Ages: a brief survey.' In R. Leighton (ed.) *Early Societies in Sicily. New Developments in Archaeological Research*, pp. 89–100. (Accordia Specialist Studies on Italy 5) London: Accordia Research Centre.

Renfrew, A.C. (1972) 'Malta and the calibrated radiocarbon chronology.' *Antiquity* 46: 199–221.

Renfrew, A.C. (1973) *Before Civilisation. The Radiocarbon Revolution and Prehistoric Europe*. London: Jonathan Cape.

Ridley, M. (1976) *The Megalithic Art of the Maltese Islands*. Poole: Dolphin Press.

Sandars, N.K. (1985) *Prehistoric Art in Europe*. Harmondsworth: Penguin.

Skeates, R. (1995) 'Animate objects: a biography of prehistoric "axe amulets" in the central Mediterranean region.' *Proceedings of the Prehistoric Society* 61: 279–301.

Stoddart, S. (in press) 'Long term dynamics of an island community: Malta 5500 BC–2000 AD.' In R. Tykot (ed.). *Social Dynamics in the Central Mediterranean*. Sheffield: Sheffield Academic Press.

Stoddart, S., Bonanno, A., Gouder, T., Malone, C. and Trump, D. (1993) 'Cult in an island society: prehistoric Malta in the Tarxien period.' *Cambridge Archaeological Journal* 3 (1): 3–19.

Tusa, S. (1992) *La Sicilia Nella Preistoria*. Palermo: Sellerio Editore.

Ugolini, L.M. (1934) *Malta. Origini Della Civiltà Mediterranea*. Roma.

Zammit, T. (1930) *Prehistoric Malta, the Tarxien Temples*. Oxford: Oxford University Press.

CHAPTER FIFTEEN

ALL THE KING'S HORSES

Assessing cognitive maps in later prehistoric Europe

COLIN RENFREW

My intention in this chapter is to take the notion of the cognitive map as a device to be used in the consideration of configurations of thought in specific contexts in prehistoric times. It is my argument that this device can be used to help us achieve some understanding of the preoccupations and interests of individuals, and groups of individuals, in particular areas. It is relevant to focus upon those areas of symbolic expression where the people creating and using symbols enjoyed some degree of freedom of expression. In such cases, it may be possible to recognise today what may be termed *cognitive constellations*, that is to say the symbolic representation of groups of associated ideas and concepts that may have been significant in forming and then in illustrating and reinforcing the ethos of the society of the day.

I shall illustrate these notions by dealing with the iconography of terrestrial transportation (wagon, chariot, horse and rider) in later European prehistory. In doing so I shall hope to demonstrate that, as we may already infer from the material finds themselves, horse riding for military purposes was not seen in Europe until the first millennium BC (the Iron Age), and that claims for the earlier military signficance of the horse are without foundation. This point has significance for European prehistory, since the myth of the nomad warrior horseman in the Bronze Age has had and continues to have a baleful influence upon Indo-European studies. These issues are relevant to the central theme of the present volume for two reasons.

Creativity is not easy to define. If by that term we mean anything more than an inclination towards self-expression, it might be defined as 'a propensity to bring about (enduring) innovation'. And innovation involves not only the formulation of the novel, but also its adoption (Renfrew 1978): innovation is a property not of the individual but of the community. For it is the community that adopts an invention, even if it may be the lone genius who formulates it. Creativity is thus a social phenomenon. It involves persuasion, teaching and communication as much as ingenuity.

It is clear that the early use of novel means of transport was, in early Europe as well as in southwest Asia, a matter of high prestige and of social status, sometimes related also to ritual concepts. The use of boats in funerary contexts in the Bronze Age Aegean, the use of carts in burials from the European Late Neolithic to the Iron Age, and the high status of cavalrymen in the later Iron Age, culminating in the convention of the equestrian portrait for Roman emperors and late medieval monarchs, all testify to this. We are dealing here with a configuration that had a powerful influence upon European prehistory and indeed upon world history. It should be noted that in many cases the power of the innovation – the new military device of the chariot or the cavalryman – was as much cognitive as technological. That is to say, the new symbol of prestige worked as much as by impressing the populace as it did through any productive capacity in its actual use. These were not necessarily efficient killing machines in terms of body counts, but if they allowed those using them to win battles, often with little bloodshed, then they were effective in the outcome. In one sense, therefore, we are indeed speaking of creativity in the field of war.

This field of study is of interest to those concerned with creativity for a second reason. We are indeed dealing here with modes of expression, with the production of visual symbols, with the development of iconography. These are fields where creativity, in a looser sense, is certainly operative. It is, in short, my view that, within the study of European prehistory, there has for too long been a separation between what is often termed 'prehistoric art' on the one hand and the material finds relating to the subject area, in this case of transportation and of war, on the other. But 'prehistoric art' is not just a matter of style. Depiction, illustration and iconography are social phenomena: the choice of subject matter may not always be easy for us to understand today, but it must in various ways have been bound up with the preoccupations and concerns of those who were creating the representations and of the wider community for whom they were working. The aim of cognitive archaeology, and here of the discussion of cognitive maps and cognitive constellations, is not merely to introduce a new jargon into the field; it is rather to bring together these various concerns so that by thinking, systematically and carefully, about the preoccupations and concerns, and indeed the mental associations (as reflected in the symbols presented) of the communities of the time, we may better understand the place of these symbolic representations within their whole use of material culture and hence within the patterns of their lives.

COGNITIVE MAPS

It is useful to assume that there exists in each human mind a perspective of the world, an interpretive framework, a cognitive map or *mappa* (Renfrew 1987a). This is a notion akin to the mental map that geographers frequently discuss, but not one restricted to the representation of spatial relationships only. To assert this is not to go down the road towards the 'empathetic' methodology of various 'post-processual' or interpretive archaeologists (Bell 1994: 305), but rather to use the individualistic method, already familiar to philosophers of science. Human beings evidently do not act simply in relation to their sense impressions. They react also in relation to their existing knowledge of the world, relating current data from sense impressions to the existing memory store. In this, of course, they do not differ from most members of the animal kingdom, and the notion of the mental or cognitive map is widely used by physiologists when analysing animal behaviour in relation to the memory store of the individual. In humans, however, the sense impressions are indeed related to the memory store, but are then further interpreted and given meaning using those concepts that find explicit symbolic form in the spoken language (Renfrew 1987a; Renfrew and Bahn 1991: 340).

This can be expressed in terms of the THINKS model, where the bubble for individual thought is modelled on the representative convention used in comic books. In Figure 15.1 we see the human individual accompanied (in his or her own mind) by this personal cognitive map, which allows the recollection of past states in the memory, and indeed the imagining of possible future states in the 'mind's eye'. For the act of recollection, how it *was* conceived as different from how it is, is closely analagous to visualising how it *would* be or *will* be (or could be). There is no doubt more to say about that. It should be noted that the experience of dreams is often not rigorous in distinguishing past from present, or present from future or from fancy.

How this mental map is structured is a matter that merits much further discussion. No doubt all that we have already said would be true of the mental maps of the anthropoid apes, whose behaviour is described by Merlin Donald (1991) as 'episodic', as well as of *H. erectus*, whose cognitition and behaviour he dubs 'mimetic'. The mental map of *H. sapiens* is clearly structured as the individual learns and grows by the concepts that are mediated and reflected by language: this justifies Donald's notion of a 'linguistic' or 'mythic' phase. The individual's mappa is structured also by experience of the man-made world, by artefacts, and in view of the active role of material culture in the formation of social relations. This is an important point (Renfrew, in press) to which we shall return below.

The relationship between the mappa of the individual and what one might loosely term the 'cognitive map' for an entire community has not

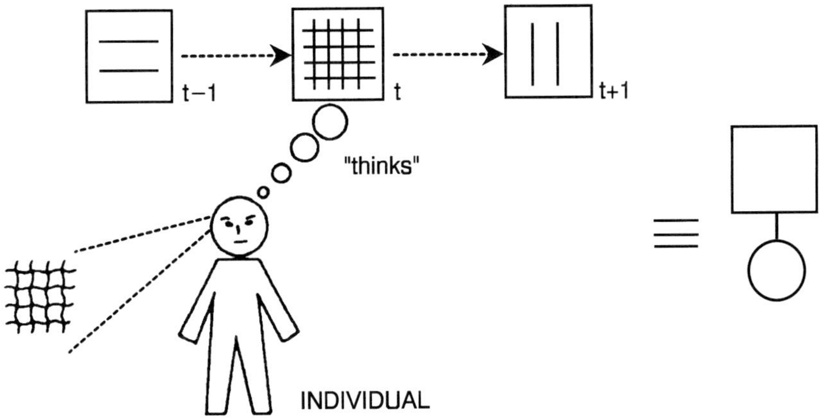

Figure 15.1 The human individual is accompanied by his or her cognitive map, represented by a square. The individual responds both to immediately perceived sense impressions and to the internalised map which includes a memory of the past (t−1) and a forecast of the world of the future (t+1).

yet been examined in detail. Yet it is clear that, in communicating with others, the individual is doing so using concepts and ideas drawn from his or her personal mappa, using the intuitive assumption that the mappa of the other person involved in the interchange is much the same. It is perhaps unnecessary to make too heavy weather of this point (although it is an important and interesting one), in the sense that when we use words to communicate with other people we tend to assume that they known what we mean – and even that they mean the same thing by a word as we do. I realise that here we touch upon the nub of linguistic philosophy, an area where no single assumption can be allowed to go unchallenged. But there is undoubtedly a strong analogy between our acceptance, in speaking our native language with another native speaker, that we understand each other, and the notion of the shared cognitive map. Within the context of this chapter it is not necessary to go further, although these points are basic to the fields of cognition and communication.

The archaeologist does, from time to time, have various, often oblique insights into the cognitive maps of earlier individuals and cultures. For although the notion of the cognitive map, as described above, is an inferred or theoretical entity, it is mirrored in our own personal experiences today, and occasionally notations indicative of aspects of the cognitive map are found. The most obvious of these are early maps, or plans apparently undertaken prior to construction work. Design layouts, undertaken in advance of some decorative scheme, are also relevant. But there is no doubt that the most informative insight into the 'mind's eye' of earlier individuals and cultures comes from pictorial representations.

FROM MAPPA TO EXTERIOR REPRESENTATION

Any depiction or representation of the world is a product of the human mind. Unless produced mechanically, as in photography, it has to go through the mind first. The features that we see in the representation are thus selected from the cognitive map or mappa of the individual who created it. Whether or not we 'read' the depiction in the same manner as its creator will have done is very much a question as to whether or not our own personal mappa is in these respects compatible with his or hers. In some respects the representation in question may be typical of its period and culture. In others it may show personal or idiosyncratic qualities (which may be another way of saying that the mappa of the creator was not identical to the shared cultural mappa of his or her contemporaries).

In this chapter it is the choice of subject matter for the depiction that is of particular interest. It is, of course, a commonplace that the painted caves of the Upper Palaeolithic show, in the main, wild animals of the day, and do so in a manner that in some respects may be termed 'naturalistic'. As such they contrast to the much more schematic depictions seen in some later periods, for instance in Spanish Levantine art, or on early pot decoration in southwest Asia, for instance the early ceramics of Susa. One of the puzzles of the European Bronze Age is that the rock art of the Scandinavian Bronze Age depicts quite a limited range of motifs (Tilley 1991) while that of Monte Bego is even more restricted. Even more puzzling in prehistoric Europe is the very limited range of contexts in which depictions are found at all. Terracotta figurines, for instance, of humans and animals mainly, are found in the Neolithic and Chalcolithic of the Aegean and southeast Europe (where the distribution extends as far west as Hungary and as far east as the Ukraine) but are rare in other parts of the continent. The stone monuments of Malta are rich in decoration and accompanied by sculptured human figures, while in general the megalithic monuments of northwestern Europe lack figured representations. Even linear decoration is limited in its distribution, although notable in specific contexts in Brittany and Ireland.

We are, of course, restricted in our view by the circumstances of preservation. Decoration in textiles, with rare exceptions, is not preserved for us from prehistoric Europe, and other fields, such as wood carving, are poorly preserved. Fortunately, from the Middle Bronze Age onwards it became increasingly common to incise or emboss figured decoration on metal objects, and in some cases to cast bronze figurines in three dimensions. During the course of the Bronze Age, the representational repertoire increases, and is abundant with the Iron Age. However, despite the limitations of preservation, and the rather puzzling figurative reticence of the populations of the European Neolithic and Early Bronze Age, some patterns do emerge.

COGNITIVE CONSTELLATIONS

In discussing the ideational background to the use of symbols and associations of symbols, especially in non-literate communities, it is relevant to draw attention to the existence of what may be termed cognitive constellations. These may be defined as dominant ideas (often relating to high status) embodied in habitual associations of symbols, given emphatic and pre-eminent form by expressive acts involving deliberate choice. At the observational level, this is a commonplace enough observation. Specific symbolic forms are often found together. In a sense, Gordon Childe's definition of the archaeological culture as 'a constantly recurring assemblage of artefacts' generalises the point. But the emphasis, when we are speaking of cognitive constellations, implies choice in expression, underlain in many cases, it may be inferred, by recurrent associations of ideas. It is not surprising that we should find artefacts pertaining to cooking in association with the animal or plant remains of the species to be cooked. That is a purely functional association, at least at first sight. But when, in Minoan Crete, we recurrently find the horned emblem, often known as the 'horns of consecration', together with the double-axe motif, there is a symbolic association that has a different significance. (It should be remarked in passing that the notion of 'functional association' can no longer be seen as unproblematic, for instance after the work of J.D. Hill on Iron Age pits [Hill 1993].)

It should be noted that any representation of the world, which therefore consists of symbols (visual) in a structured configuration, may constitute a cognitive constellation. But such constellations are not restricted to pictorial representations in two dimensions. The essential ingredient is the exercise of choice: the symbols whose association forms the constellation are the product of selection by the individual (who may in this be constrained by the conventions of the community, in themselves theoretically a matter of choice, even if the choice is regulated by convention).

Choice of expression

There are many symbolic fields where the individual exercises a wide range of choice (although, as noted, the choice may be regulated by the norms applying within the community). Among these are:

Dress Aspects of dress may be regulated by gender, age, professional occupation, status, etc. In this way some items of dress by convention carry coded meaning about these matters, while others offer scope for personal expression by the individual.

Monuments Structures erected specifically to carry meaning, and to act as memorials, being in themselves symbols are often rich in further symbolism.

Creation of figurative iconic complex As implied above, any figurative representation of the world offers a huge range of choice, where symbols are arranged in a way that carries particular meaning for the individual.
Decoration Any decorative scheme offers a wide range of choices, even if the size, shape and material of the artefact to be decorated imposes some restrictions. Even when the decorative schema is a non-figurative one, there are still many choices to be made, many of which may be symbolically significant.
Burial Burial offers the inescapable necessity of making a number of choices, among which is the selection of appropriate gravegoods to accompany the deceased. It is one area where cognitive constellations are particularly evident to the archaeologist.

Choice in burial

The formal burial of a deceased individual involves a number of choices, and there is, of course, an extensive archaeological literature on this matter (e.g. Chapman *et al.* 1981; O'Shea 1984). Decisions have to be made (although many are determined, as it were, by the prevailing conventions) about: time and place of burial, type (inhumation, cremation, excarnation, etc.), form (burial or exposure, nature of monument), etc. If the deceased is clothed, the conventions of dress involve choices that may relate to those of daily life or may be governed by special mortuary conventions.

In particular, the determination of gravegoods offers particular scope for choice, which may convey the image or social persona that those arranging the burial wish to impart in relation to the deceased. If, for example, the deceased was an important leader and powerful warrior, this may be conveyed by associations of gravegoods designed to convey this message, as no doubt in many cases they did in use during the life of the deceased. The 'princely' burials of Late Halstatt and Early La Tène France and Germany offer an excellent example (e.g. Wells 1980; Figure 15.2). All these instances that involve choice inevitably lead those responsible for the choice to betray their own preoccupations and interests. To distinguish the personal interest of the individual responsible (whether artist, designer or organiser – and, in the case of a burial, of the deceased person) from the constraints upon choice exercised by the prevailing social conventions may not be an easy matter, and the issue certainly merits further consideration. Some aspects of the question, but not all, have been touched on by Hill and Gunn (1977) in *The Individual in Prehistory*. To what extent the social conventions and norms of a particular culture are to be taken as reflecting the 'preoccupations and interests' of the individuals who participate in it is a matter for discussion also. But clearly these conventions and norms are the product, over the years, of successive choices that have, for whatever reason, become adopted as standards.

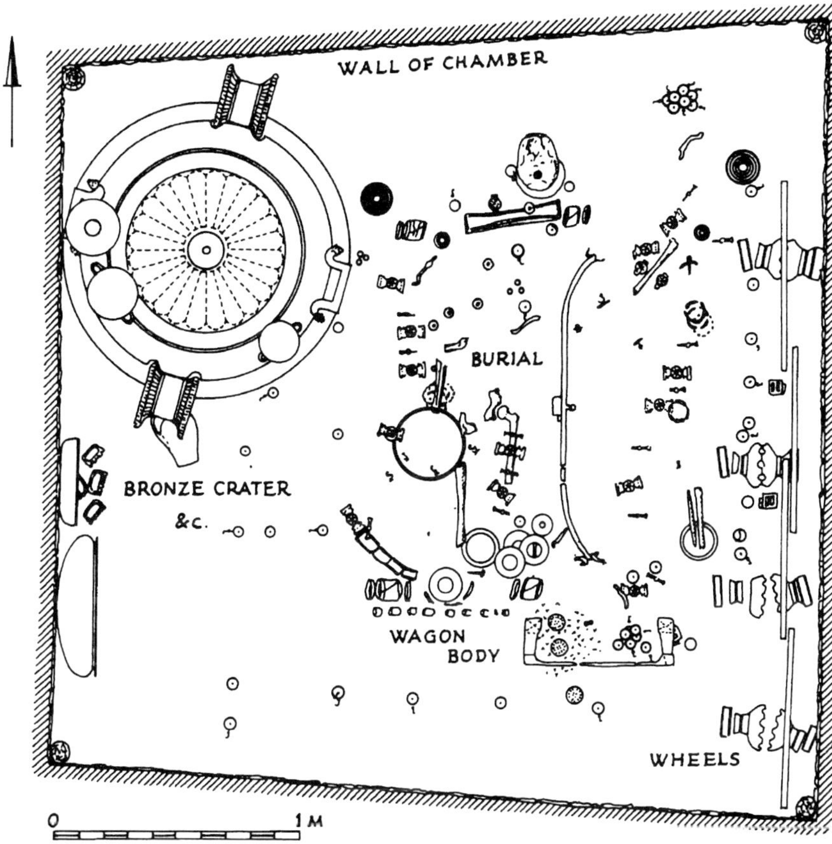

Figure 15.2 Choice in burial and the construction of social identity: the rich burial of the 'princess' of Vix in the fifth century BC (after Piggott 1983)

THE HORSE IN EUROPEAN PREHISTORY

The theme of the use of the horse in European prehistory is a familiar one, partly because since the time of Gordon Childe and particularly of Marija Gimbutas the theory has been propounded that mounted warrior nomads from the steppe lands north of the Black Sea invaded central and western Europe at the beginning of the Bronze Age, bringing with them the proto-Indo-European language. This was in effect, it is argued, the 'coming of the Indo-Europeans', a significant development for European prehistory. This is not the place for a discussion of linguistic matters, some of which I have touched on elsewhere (Renfrew 1987b, see also Mallory 1989). Here I want to discuss the evidence for the riding of the horse for military purposes, and to show that the myth of the mounted warrior horseman at the beginning of the Bronze Age is without foundation. It finds clear expression, for instance, in the words of Diamond (1991: 244): 'With horse domestication the steppe peoples became the first to put together the economic and military package that came to dominate the world for the next 5000 years.' Here Diamond is falling into serious confusion, equating the very tenuous evidence found by Anthony (1986) at the Ukrainian site of Dereivka, indicating some tooth wear indicative of sporadic use of the bit perhaps around 5000 BC (and hence perhaps some riding of horses in the course of herd management), with the military use of the horse. But there is no evidence for such use in Europe until around 1000 BC. And without such military possibilities, the whole explanatory basis for the supposed 'kurgan' invasions at the beginning of the Bronze Age disappears.

This then is the background to the brief survey that follows. It serves to illustrate the earlier discussion because systematic use is made here not only of direct evidence or transport resources, including horses, but also of depictions of these. It will be seen that first carts, and then chariots, drawn by horses, and then indeed warriors on horseback held a special significance in European prehistory. They each formed, in their day, part of a cognitive constellation relating to the prestige of an elite minority and to the appropriate burial conventions accorded to that minority. For this reason, no doubt, we see them in a number of representations and depictions. It is an important part of my argument that, given that these things were indeed components of cognitive constellations that were selected for depiction and representation, their occurrence and *their non-occurrence* in such depictions is of significance. It will be argued, for instance, that since warriors on horseback are depicted with some regularity in the first millennium BC but are lacking from earlier depictions, while warriors in horse-drawn chariots are seen from the middle of the second millennium BC, warriors on horseback did not form part of a significant cognitive constellation during the second millennium. It will further be argued that, since the warrior on horseback, like the horse-drawn chariot before it, had such a considerable impact when it was indeed introduced, the mounted warrior was not in fact a significant feature of society prior to its early representation around 1000 BC.

In this way, the evidence from representations and depictions can legitimately be introduced into the discussion alongside the direct evidence, derived from archaeological excavation, for the use of these transport facilities. In undertaking such a survey, I have been greatly helped by the wealth of documentary evidence assembled by Gordon Childe, T.G.E. Powell, Sinclair Hood, Stuart Piggott, Littauer and Crouwel and others, and by recent work by Ukrainian and Russian scholars (e.g. Kuzmina 1994). When considered systematically as an indication of deliberate symbolic representation rather than as a simple reflection of knowledge of horsemanship, a reasonably clear picture emerges. To discern it, one has to see the military use of the horse as part of the mindset of the community in question – as part of a cognitive constellation – rather than as something that follows immediately upon the availability of the horse.

THE HORSE AS A FOOD RESOURCE: EARLY 'DOMESTICATION'

There are no indications of the use of the horse in western Europe for purposes of traction before the second millennium BC.

The horse was, of course, found throughout much of Europe in Upper Palaeolithic times. Indeed the suggestion has been made (Bahn 1978) that there are indications from that time that it was intensively exploited, and its movements constrained by means of enclosures. That the horse survived in several parts of Europe into Neolithic times and beyond is indicated by a series of finds. The well-known decorated bones with oculus motifs from Los Millares and other Late Neolithic sites in Iberia are often horse bones.

There are finds of horses, possibly domesticated, from the Beaker period at Newgrange (van Wijngaarden-Bakker 1986). It is widely assumed that these horses were a food resource, as at the important site of Dereivka in the Sredny Stog culture of the Ukraine, but they may also have been used for traction, perhaps supplementing the use of oxen for this purpose. Indeed it is not impossible that the intensive use of the horse as a food resource started in the Ukraine, and that this was the source of the horses utilised in western Europe. It is to be hoped that genetic studies will soon clarify the question of the original habitat of the wild horses that were exploited in this way. Certainly the horse was well known as a wild animal in several areas of Europe before its more intensive exploitation, so that local domestication remains a possibility. There are also reports of finds from Anatolia and western Iran.

The exploitation of the horse at Dereivka and related sites is so intense (more than 60 per cent of the faunal remains were horse bones) that they must certainly have been herded, and their breeding may also have been controlled. In this sense, they were 'domestic' animals. But the 'domestication' of the horse involves so many factors (Levine 1990) that caution must be exercised in employing the term too readily. Anthony (Anthony and Brown 1991) has made the claim from the wear studied on a single horse tooth from Dereivka that as early as 4000 BC horses were induced to wear mouth bits of some kind, perhaps for purposes of riding. There is nothing inherently improbable in such a suggestion – some measure of riding of particularly docile animals would be convenient for herd management – but the context of the tooth examined by Anthony has been questioned at a recent conference in Kiev and the later date of 3000 BC has been suggested (Levine, pers. comm.). Further work by Anthony has not revealed further examples.

It is worth noting at this point that these indications of horses in several parts of Europe in the Neolithic period undermine one old argument pointing to the steppe lands as the 'homeland' of the Proto-Indo-Europeans. It was argued that the existence of a word for horse (in fact there are several such words) in early Indo-European speech was a secure indication that the Indo-European dispersals must have occurred no earlier than the Early Bronze Age, since the horse was not domesticated before that time. But the question of the domestication of the horse is rather a different one from that of its nomenclature.

Lichardus (1980) has claimed that pierced and pointed pieces of antler found in a number of European locations at the end of the Neolithic period, may have served as cheek pieces for horses. And it is the case that the examples he cites resemble finds from Dereivka. It is more likely, however, that if these were indeed associated with horses, they had a role in relation to traction rather than riding. For in shape they resemble somewhat the bone cheek pieces of the Fuzesabony culture of the Hungarian Middle Bronze Age. These are probably to be associated with the use of horses to pull the chariots with two spoked wheels that we are about to discuss. There are only rare indications that horses were sometimes used to pull carts with solid wheels, although the equids depicted on the Royal Standard from Ur are doing just that. But the going was probably more difficult on the muddy terrain of temperate Europe.

Unless I am mistaken, we find no depictions of horses in Europe, following the notable art of the Upper Palaeolithic, until the time of the Shaft Graves of Mycenae, around 1600 BC, when the horse was used to pull the chariot with spoked wheels (Figure 15.3). Although the horse was used as a food resource before that time, it is difficult to imagine that it was 'domesticated' in any wider sense until the introduction of the chariot. Nor did it form part of any notable cognitive constellation of the kind discussed above, or at least not to the extent of inspiring symbolic representations.

CARTS BEFORE HORSES

In the Late Neolithic period of Europe we find clear indications of the use of wheeled vehicles. The wheels themselves are of wood, solid wood, and generally formed of three pieces, fitted together. The earliest finds from Europe are probably the rare occurrences of wheels themselves (van der Waals 1964). One example, made of a single piece of wood with integral nave, comes from De Eese in Holland, and was found near the remains of a wooden trackway of corduroy construction preserved in a peat bog, dated to *c.*2100 BC in radiocarbon years. Another comes from near Zurich. Fortunately we can gain some idea of the completed cart from several sources. The nearest-to-hand is the series of cart models

Figure 15.3 Princely pursuits: the horse and chariot, first seen in Europe in the mid-second millennium BC, complete with the finery of well-armed chieftains of the Mycenaean Shaft Graves (after Schliemann 1880)

from the Baden culture of Hungary and neighbouring lands, of which the best-known example comes from Budakálasz (Figure 15.4).

Complete carts, with tripartite wheels, have been found in burials of the Pit Grave culture in Armenia (Häusler 1981), and from later contexts under burial mounds at Trialeti in Georgia and at other sites. The wagons with disc wheels found in Georgia and Armenia seem in general to have been pulled by oxen. Certainly they are seen pulled by oxen in the rock engravings from the second millennium BC at Syunik in Armenia (Piggott 1983: 78).

Burials of paired oxen occur from sites of the Baden culture in Hungary, for instance at Alsonomedi (Behrens 1963), and Banner (1956: 207)

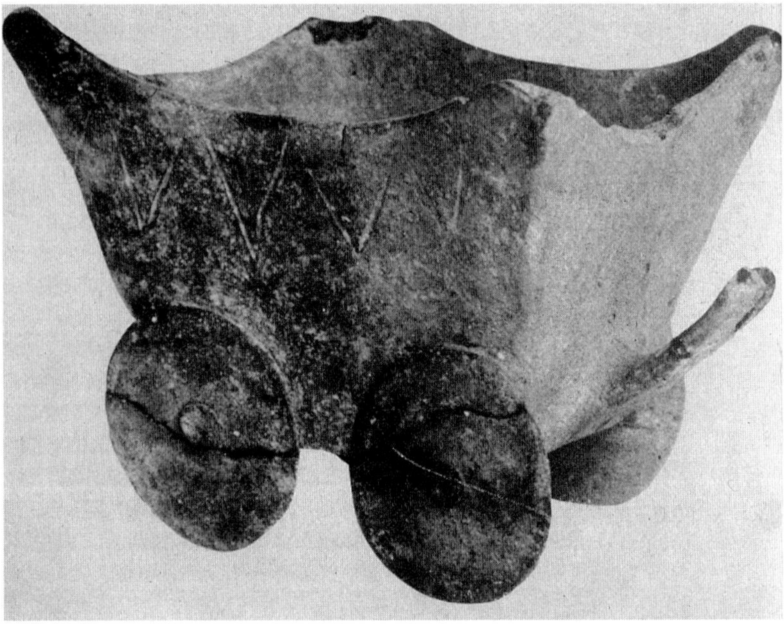

Figure 15.4 Pottery cart from Budakalász (Baden culture) *c.* 3000 BC (after Piggott 1983)

suggested that these were draught oxen. It may be claimed that the four-wheeled cart pulled by paired oxen achieved considerable symbolic importance over a wide area, from Georgia and Armenia to Val Camonica. It clearly formed part of a cognitive constellation, relating to burial, and the choice of the cart not only for actual burials but for depiction in models and in rock engravings is an indication of this.

Certainly there are numerous indications that the four-wheeled cart with solid wheels and drawn by two oxen had an impact upon contemporary consciousness. One of the few incised pots of the TRB culture to show a figurative design, the pottery cup from Bronocice in Poland (Milisauskas and Kruk 1977), indicates a four-wheeled cart (Figure 15.5), and although oxen are not depicted there, their use is supported by the remarkable find of a pair of oxen in copper from the TRB culture at Bytyn in Poland (Figure 15.6; Piggott 1983: 42).

Rock engravings of ox-drawn wagons are seen at Val Camonica and generally dated to the third millennium BC.

We thus see clear indications that the first wheeled vehicles in Europe, which had solid disc wheels, were pulled by oxen as early as 2500 BC in radiocarbon years, and hence well before 3000 BC in calendar years. The horse simply does not come into the picture yet, so far as transport is concerned. Moreover, the local or multiple invention of the solid wheel

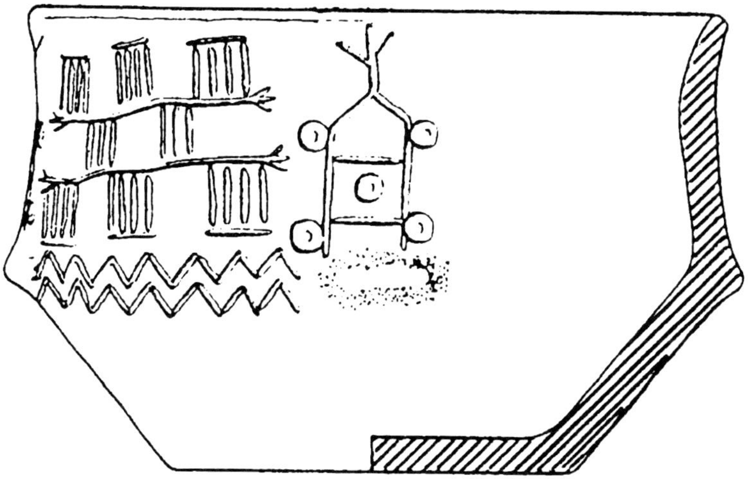

Figure 15.5 Incision on TRB pot of Bronocice showing four-wheeled cart, Poland, fourth millennium BC (after Piggott 1983)

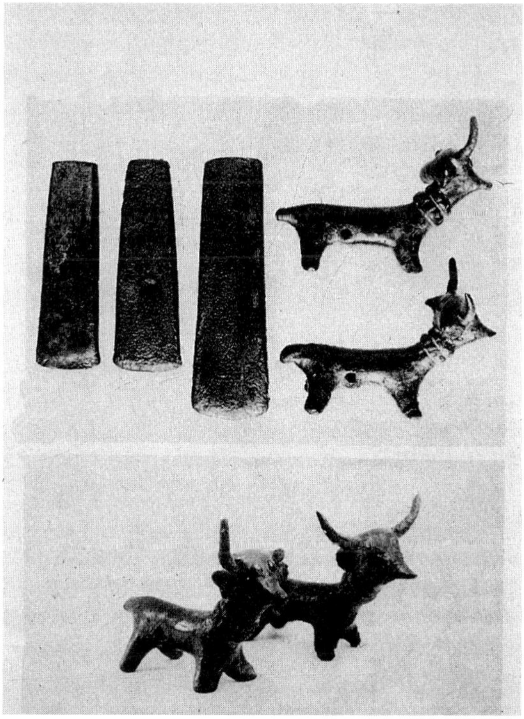

Figure 15.6 Pair of copper oxen from Bytyn (TRB culture) third millennium BC (after Piggott 1983)

still remains a possibility. Childe (1951) laid emphasis upon the tripartite wheel as representing a special manufacturing technique that may have had a single point of origin. But it is clear now that the solid disc wheel is an earlier form in temperate Europe, and that issue remains an open one. As Childe concluded, at the end of his pioneering survey:

> Everywhere within the period here considered the vehicles were drawn by beasts, yoked on either side of a central pole. In other words the method of traction adopted for carts and waggons was that approved for ploughs.
>
> Oxen are among the earliest animals known to have been thus yoked to vehicles in Mesopotamia, India, Georgia and Sweden. But there and elsewhere oxen had probably been drawing ploughs before they were harnessed to carts or waggons. The latter merely replaced the plough share at the end of the pole.
>
> (Childe 1951: 193)

The functional association between ox draught in relation to the plough and to the wagon perhaps makes more likely an origin in temperate Europe, since plough agriculture was perhaps less likely in the pastoral, horse-rearing economy of the steppes.

THE CHARIOT HORIZON

Perhaps the term horizon is a dangerous one to use in archaeology these days. It is deliberately employed here to emphasise that within the space of just a few centuries, over a very wide area, the horse makes its appearance pulling the two-wheeled chariot with spoked wheels. Horses were used also from that time onwards to pull the four-wheeled carts whose spoked wheels made them much lighter than the carts with disc wheels discussed above.

Already in Mesopotamia in the third millennium BC, two- and four-wheeled vehicles with solid wheels were used in burials, and in contexts indicative of prestige. But it is clear that they were displaced in this respect by the horse and chariot when these made their appearance.

Of course we see them for the first time in Europe in the stelai of the Shaft Graves of Mycenae from about 1600 BC. Chariots are of frequent occurrence in Minoan and Mycenaean art from around that time. They occur a couple of centuries earlier in Egypt, from the Hyksos period (Schulman 1995), and from the same period in the Near East (Littauer and Crouwel 1979; Dalley 1995).

Kuzmina (1994) emphasises the importance of horse breeding and chariots in the steppe lands from the origins of the Andronovo culture, with rich warrior burials in the Urals and Northern Kazakhstan in the sixteenth century BC (Figure 15.7). More recent finds in the same area (Piggott 1983: 91; Anthony and Vinogradov 1995), supported now by radiocarbon dates from the site of Krivoe Ozero, indicate that chariot burials began there as early as 2000 BC, and the consistent pattern of other finds

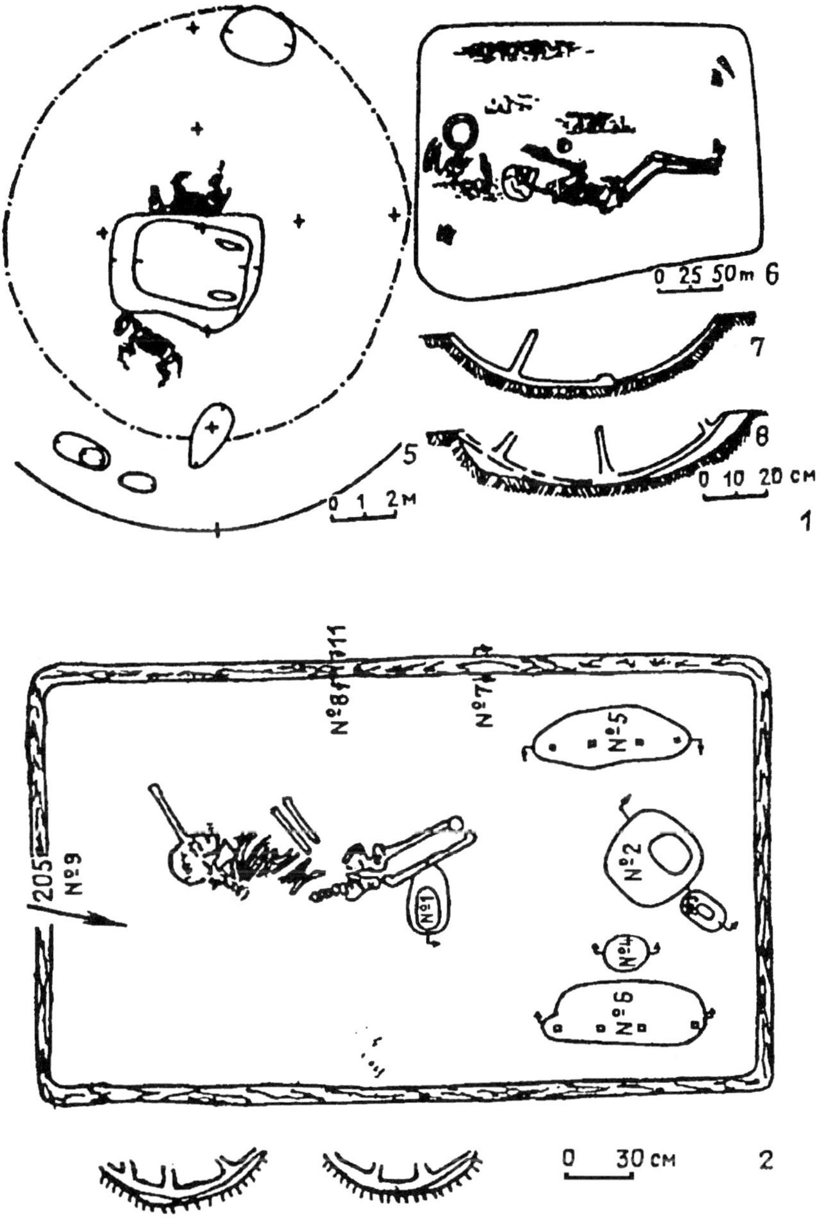

Figure 15.7 Andronovo culture graves with chariots and the imprints of spoked wheels *c.* 1550 BC (after Kuzmina 1994)

indicates that there is no doubt that they were drawn by horses.

The picture outlined by Kuzmina (1994) is a persuasive one, that the use of the chariot spread rapidly on the steppe lands of Russia and south to the Iranian plateau. Our earliest knowledge of Indo-European speech in that area comes, appropriately enough, in tablets of the thirteenth century BC dealing with the training of horses (for chariots) found at the Hittite capital of Bogazköy. The text is in the Hittite language, but written by Kikkuli, a Hurrian from the state of Mitanni. It contains technical terms in horsemanship that are clearly derived from a language related to Indo-Iranian. It is clear that the occupants of Egypt and Mesopotamia felt less at home with horses than their contemporaries on the Iranian plateau. So it is entirely plausible that, as Kuzmina argues, developments in horse-breeding and the development of the spoked wheel at the beginning of the second millennium BC on the steppe lands led for the first time to a military vehicle, the chariot, that was of real significance in battle. It is so depicted in Egyptian reliefs in the fourteenth century BC and it is so portrayed also in the Hymns of the Rig Veda which may date from about the same time.

But while the arrival of the war chariot, pulled by the first horses used in warfare, may well have favoured the southward expansion of a more mobile nomad pastoral economy towards the Indian sub-continent (taking with it early Indo-Iranian languages) its significance in Europe and the Near East was not quite so great. Few writers today would dispute that the Hittites were established in Anatolia before that time, nor that the Greek language was established in Greece prior to the introduction of the chariot in about 1600 BC.

That being said, the impact in Europe was considerable. Model spoked wheels are found in a number of contexts, and chariots pulled by horses clearly formed part of a cognitive constellation, which in some ways took over the significance of the wagon drawn by oxen. The incisions on the Kivik cist (Figure 15.8) and the Trundholm cart (Figure 15.9; admittedly of four wheels, but they are spoked wheels, and drawn by a horse) are clearly of considerable symbolic significance, and they are generally thought to be only a century or two later than the stelai of Mycenae.

It should be stressed, however, that there is no indication of horse riding at this early time.

THE HORSEMAN HORIZON

The inception of horse riding in the steppe lands is dated by Kuzmina (1994) to the twelfth century BC (Figure 15.10). But we see very early indications dating from the same time, or earlier, in the Near East and in Greece. The Assyrian palace reliefs and some of the Egyptian royal reliefs already in the thirteenth century BC show horses being ridden away from the scene of battle by the defeated enemy (Figure 15.11).

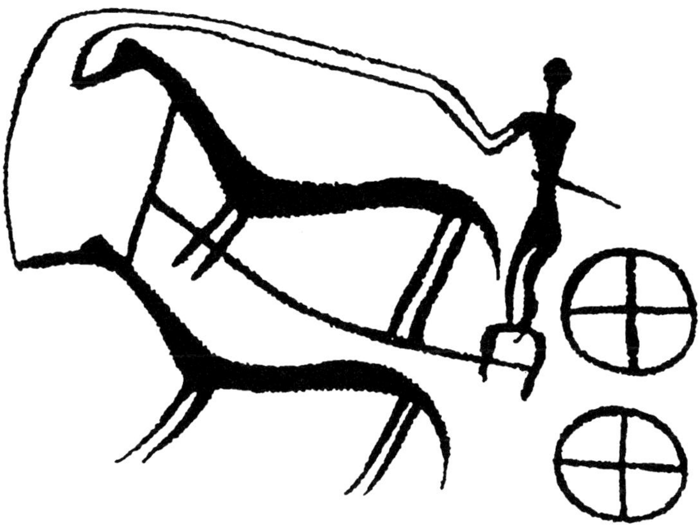

Figure 15.8 Chariot scene from the Kivik cist, Sweden *c.* 1300 BC

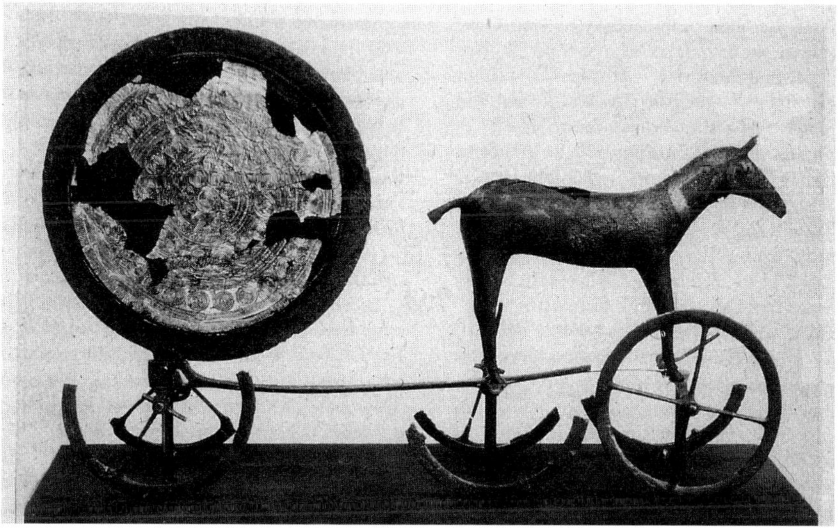

Figure 15.9 The Trundholm disc: on a horse-drawn cart with spoked wheels, Denmark, *c.* 1300 BC

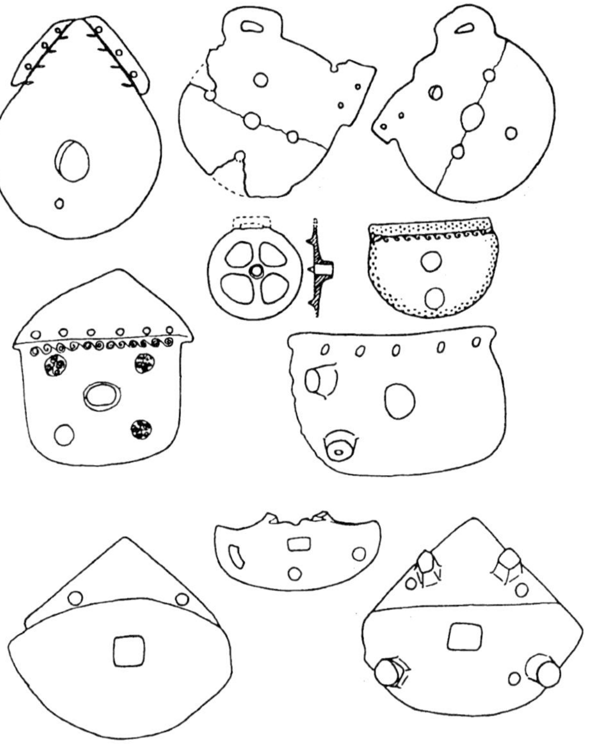

Figure 15.10 Psalia (cheek pieces) of the steppes from the 12th century BC. The circular spoked piece is from Mycenaean Greece (after Kuzmina 1994)

They are, however, ridden on the withers rather than on the thorax. It seems clear that these may be chariot horses that are being used in this way. The Egyptian sign for 'commander of horsemen' in the middle of the Eighteenth Dynasty (*c.*1450 BC) seems to represent a mounted figure (Schulman 1995), and it is indeed likely that the grooms responsible for chariot horses would occasionally mount them. It is possible, then, that horse riding might have evolved independently anywhere that chariotry was taken seriously. But the skills of horsemanship may have developed rather earlier on the steppes, and it is perfectly possible that horse riding was introduced from there in the centuries after 1200 BC.

Among the earliest depictions of ridden horses, ridden that is to say for military purposes and by an armed warrior, is the terracotta figurine from Mycenae, published by Hood (1953) and dating from the thirteenth century BC (Figure 15.12). A relief slab from the Neo-Hittite period at Tell Halaf in Syria is probably a little later (Figure 15.13). The earliest impact of horse riding upon temperate Europe was assigned by Childe (1950: 229) to the 'Thrako-Kimmerians' in about the eighth century BC,

Figure 15.11 Egypt: fleeing rider (from the Relief of Horemheb, Sakkara: XVIIIth Dynasty, reign fo Tutankhamon 1332–1323 BC)

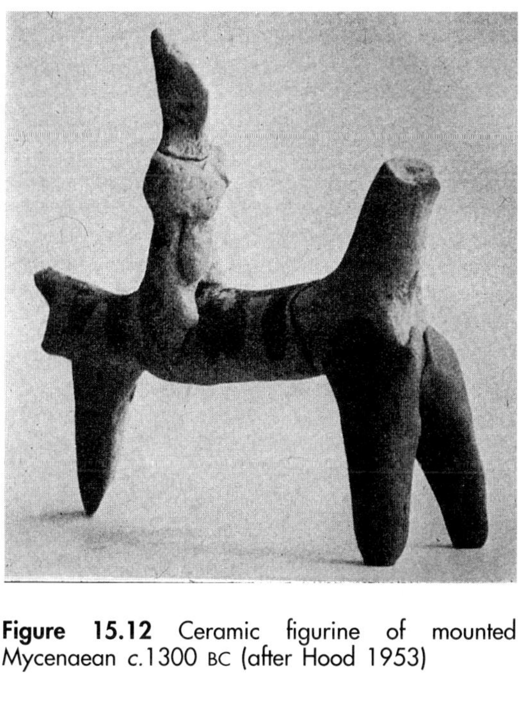

Figure 15.12 Ceramic figurine of mounted Mycenaean *c.*1300 BC (after Hood 1953)

Figure 15.13 Horseman from Tell Halaf, Syria, c. 9th century BC (after Bossert 1951)

and bronze horse harness including bronze bits are found from this time onwards.

The first depictions of ridden horses in temperate Europe occur, interestingly enough, in association with spoked-wheeled carts. There is an engraving upon a pot of the Hallstatt C or D period from Sopron in Hungary that shows a mounted figure going before a horse-drawn cart, and the celebrated bronze model cart from Strettweg in Austria, which is of the Halstatt C period (*c.* sixth century BC), also depicts a mounted warrior. From this time on there are frequent images of mounted warriors in contexts where there are depictions or figurations. This is particularly true of the situla art of Slovenia, and horsemen are seen also (although not very frequently) in Swedish rock art. There are occasional depictions in the Villanova culture of Italy, and frequent representations in the tomb paintings of the Etruscans. Of course, the Scythian tombs of the steppes, and tombs in the Greek colonies adjacent to them, produce abundant representations of armed warriors; and with the Scythians we have the graphic descriptions of Herodotus, which give the earliest detailed insights into the nomadic way of life. From about the fifth century BC we see archers able to manoeuvre freely on horseback, releasing that rearward-facing 'Parthian shot' mentioned by the classical writers.

Prestige: riding in state

There are just a few indications from a variety of contexts that to ride a horse may not have been as prestigious an undertaking as to be driven in a chariot. Powell (1971: 2) draws attention to the breach of regal behaviour by the King of Mari, Zimri Lim, who indulged in horse riding while visiting his nomad subjects, and Alexander the Great is one of the first rulers shown leading his troops into battle on horseback, at the Battle of Issos, on the celebrated mosaic in the Naples Museum. But the Assyrian monarch Ashurbanipal is already seen shooting lions from the saddle in one of the reliefs from his palace at Nineveh. Certainly the four-wheeled cart was the preferred vehicle for burial in Hallstatt times, and the two-wheeled chariot in the La Tène period. And although Caesar encountered cavalry in Gaul (he makes no mention there of chariots), in the Old Irish folk tales it is the chariot that figures, without any mention of cavalry (Piggott 1983: 236). As noted earlier, it is the complete symbolic system, the cognitive constellation, that determines what is represented and even what is used. The availability of the necessary technology is not the determining factor. By the time of the Roman emperors, however, to ride on horseback was an appropriate mode of transport for a ruler. And the equestrian statue in bronze of Marcus Aurelius survived throughout the medieval period on the Capitoline Hill in Rome, where it is still to be seen today.

IMAGE AND REALITY

It is clear from the foregoing that *there is no evidence of warrior horsemen prior to 1300 BC anywhere in the Mediterranean, in Eurasia or beyond.* Horses were an important food resource in the steppe lands from the end of the Pleistocene, and were controlled, and in that sense 'domesticated', in the Ukraine from *c.*5000 BC, and used as a food resource in western Europe from *c.*2000 BC. But in terms of transport, the first major innovation was the four-wheeled cart, with solid wheels, drawn by oxen. In the context of this chapter, what interests us is not simply that carts were used for ceremonial burials as early as the Baden culture in Hungary; it is that model carts of terracotta were produced at this time (when representation was rare, and mainly restricted to the human form). Among the earliest representations made in metal from Europe of any kind are the cattle from the TRB culture of Poland. We have here indications of a particular focus of interest. Among the earliest rock engravings from Val Camonica are representations of carts drawn by oxen. There is a conjuncture between cart, oxen and burial: perhaps too simple to be hailed as a 'cognitive constellation', it persists, in a different form, into the Iron Age. By then the wheels are spoked, and the traction is sometimes by the horse. But by the time of the rich spoked cart burials of the Hallstatt Iron

Age, we can certainly speak of a cognitive constellation, where the cart, serving no doubt as a hearse, plays an important role in the ceremonial surrounding chiefly burial. The chariot horizon brings a fresh impetus. The cart was rarely a front-line vehicle for warfare (although it should be noted that in Early Dynastic Mesopotamia, several centuries before the introduction of the chariot, we do see depictions of carts in military contexts). The chariot was the ideal mobile platform for the prince, whether in warfare or the hunt, and it is depicted thus from the Shaft Graves of Mycenae around 1600 BC to the Assyrian palace reliefs of *c.*800 BC. In central and northern Europe we find special depictions of the spoked wheel and the horse-drawn chariot in various special contexts, including rock engravings. The Kivik cist in Sweden represents a funerary context. The Trundholm disc, a golden sun disc on a cart with spoked wheels pulled by horses, is part of a related cognitive constellation, to which the four-wheeled cart was assimilated when spoked wheels became available, permitting traction by horses. Although the horse and rider make their first appearance on the steppe in the twelfth century BC, and in the Mycenaean world soon after, we do not see consistent depictions of mounted warriors until about the eighth century BC (with the Assyrian palace reliefs) and then in much of Europe from the seventh and sixth centuries. The horse and rider motif becomes a very familiar one: in Thrace it came to be commonly used for grave markers, from which the horse and rider motif of the early Christian St Demetrios derives, and no doubt (with the addition of a dragon) the familiar image of St George. Although demoted in sanctity by the Roman Catholic Church, St George still survives in the English coinage as the principal motif of the golden sovereign and the silver crown, and the equestrian figures of the Great George is still the principal insignia of the Royal and Most Noble Order of the Garter. The cognitive constellations may be restructured and vary in their associations, but, at least in this respect, the iconography is remarkably consistent.

ACKNOWLEDGEMENTS

The permission of the following organisations to reproduce plates is gratefully acknowledged: Thames and Hudson Ltd (Figures 15.4 and 15.5); National Museum of Denmark (Figure 15.9); Bologna, Museo Civico Archeologico, Palagi (Nizzoli) Collection KS 1889 (Figure 15.11); British School at Athens (Figure 15.12).

REFERENCES

Anthony, D.W. (1986) 'The "Kurgan culture", Indo-European origins and the domestication of the horse: a reconsideration.' *Current Anthropology* 27: 291–313.

Anthony, D.W. and Brown, D. (1991) 'The origins of horseback riding.' *Antiquity* 65: 22–38.

Anthony, D.W. and Vinogradov, N.B. (1995) 'Birth of the chariot.' *Archaeology* 48: 36–41.

Bahn, P. (1978) 'The "unacceptable face" of the West European Upper Palaeolithic.' *Antiquity* 52: 183–192.

Banner, J. (1956) *Die Péceler Kultur* (Archaeologia Hungarica 35), Budapest: Hungarian Academy of Sciences.

Behrens, H. (1963) 'Die Rindrskelettfunde der Péceler Kultur.' *Acta Archaeologica Hungarica* 15: 33–36.

Bell, J.A. (1994) *Reconstructing Prehistory: Scientific Method in Archaeology*. Philadelphia: Temple University Press.

Chapman, R., Kinnes, I. and Randsborg, K. (eds) (1981) *The Archaeology of Death*. Cambridge: Cambridge University Press.

Childe, V.G. (1950) *Prehistoric Migrations in Europe*. London: Kegan Paul, Trench, Truber & Co.

Childe, V.G. (1951) 'The first waggons and carts – from the Tigris to the Severn.' *Proceedings of the Prehistoric Society* 17: 177–194.

Dalley, S. (1995) 'Ancient Mesopotamian military organisation.' In J.M. Sasson (ed.) *Civilisations of the Ancient Near East*, pp. 289–300. New York: Charles Scribner's Sons.

Diamond, J. (1991) *The Rise and Fall of the Third Chimpanzee*. London: Vintage.

Donald, M. (1991) *Origins of the Modern Mind*. Cambridge, MA: Harvard University Press.

Häusler, A. (1981) 'Zur ältesten Geschichte von Rad und Wagen im nordpontischen Raum.' *Ethnographisch-archäologische Zeitschrift* 224: 581–647.

Hill, J.D. (1993) 'Can we recognise a different European past? A contrastive archaeology of later prehistoric settlements in southern England.' *Journal of European Archaeology* 1: 57–76.

Hill, J.N. and Gunn, J. (eds) (1977) *The Individual in Prehistory*. New York: Academic Press.

Hood, M.S.F. (1953) 'A Mycenaean cavalryman.' In A.J.B. Wace *et al*. 'Mycenae 1939–52', *Annual of the British School of Archaeology at Athens* 48: 84–93.

Kuzmina, E.E. (1994) *Where Had the Indo-Aryans Come from? The Material Culture of the Andronovo Tribes and the Origins of the Indo-Aryans*. Moscow: Russian Academy of Sciences.

Levine, M. (1990) 'Dereivka and the problem of horse domestication.' *Antiquity* 64: 627–640.

Lichardus, J. (1980) 'Zur Funktion der Geweihspitzen des Typus Ostdorf.' *Germania* 58: 1–24.

Littauer, M.A. and Crouwel, J. (1979) *Wheeled Vehicles and Ridden Animals in the Ancient Near East*. Leiden: Brill.

Mallory, J.P. (1989) *In Search of the Indo-Europeans*. London: Thames & Hudson.

Milisauskas, S. and Kruk, J. (1977) 'Archaeological excavations at the Funnel Beaker (TRB) site of Bronocice.' *Archaelogia Polona* 18: 205–228.

O'Shea, J. (1984) *Mortuary Variability, an Archaeological Investigation*. New York: Academic Press.

Piggott, S. (1965) *Ancient Europe*. Edinburgh: Edinburgh University Press.

Piggott, S. (1983) *The Earliest Wheeled Transport, from the Atlantic to the Caspian Sea*. London: Thames & Hudson.

Powell, T.G.E. (1971) 'The introduction of horseriding to temperate Europe: a contributory note.' *Proceedings of the Prehistoric Society* 37 (Part 2): 1–14.

Renfrew, C. (1978) 'The anatomy of innovation.' In D. Green, C. Haselgrove and M. Spriggs (eds) *Social Organisation and Settlement*, pp. 89–117. Oxford: British Archaeological Reports International Series 47.

Renfrew, C. (1987a) 'Problems in the modelling of socio-cultural systems.' *European Journal of Operational Research* 30: 179–192.

Renfrew, C. (1987b) *Archaeology and Language, the Puzzle of Indo-European Origins*. London: Jonathan Cape.

Renfrew, C. (in press) 'Mind and matter: cognitive archaeology and external symbolic storage.' Paper presented to the Symposium on External Symbolic Storage, McDonald Institute for Archaeological Research, September 1996.

Renfrew, C. and Bahn, P. (1991) *Archaeology, Theories, Methods and Practice*. London: Thames & Hudson.

Schliemann, H. (1880) *Mycenae*. London.

Schulman, A.R. (1995) 'Military organisation in pharaonic Egypt.' In J.M. Sasson (ed.) *Civilisations of the Ancient Near East*, pp. 289–300. New York: Charles Scribner's Sons.

Tilley, C. (1991) *Material Culture and Text: The Art of Ambiguity*. London: Routledge.

van Wijngaarden-Bakker, L.H. (1986) 'The animal remains from the beaker settlement at Newgrange, Co. Meath.' *Proceedings of the Royal Irish Academy* 86: 17–111.

Waals, J.D. van der (1964) 'Neolithic disc wheels in the Netherlands.' *Palaeohistoria* 10: 103–146.

Wells, P.S. (1980) *Culture Contact and Culture Change, Early Iron Age Central Europe and the Mediterranean World*. Cambridge: Cambridge University Press.

INDEX